REMOTE SENSING AND GEOGRAPHIC INFORMATION SYSTEMS:

Geological mapping, mineral exploration and mining

THE ELLIS HORWOOD LIBRARY OF SPACE
SCIENCE AND SPACE TECHNOLOGY

SERIES IN SPACE TECHNOLOGY

Series Editors: JOHN W. MASON, B.Sc. Ph.D.
D. R. SLOGGETT, M.Sc. Ph.D.

METALLURGICAL ASSESSMENT OF SPACECRAFT PARTS AND MATERIALS
BARRIE D. DUNN, Head of Metallic Materials and Processes Section, ESA-ESTEC
SOLAR POWER SATELLITES: The Emerging Energy Option
Editors: P. E. GLASER, Vice President, Arthur D. Little Inc., USA, F. P. DAVIDSON, Coordinator,
Macro-Engineering Research Group, Massachusetts Institute of Technology, USA, and K. I. CSIGI,
Principal Consultant, ERIC International, USA
LIVING AND WORKING IN SPACE: Human Behaviour, Culture and Organization
PHILIP R. HARRIS, President, Harris International Ltd, and Senior Scientist, NETROLOGIC Inc.
SATELLITE IMAGING INSTRUMENTS: Principles, Technologies and Operational Systems
C. B. PEASE, Royal Aerospace Establishment, Farnborough, Hampshire
AUTOMATION IN SPACE: Vol. 1: Expert Systems in Space; Vol. 2 Robots in Space
DAVID R. SLOGGETT, Marcol Group Ltd, Woking, Surrey
SATELLITE DATA: Processing, Archiving and Dissemination
Vol. I: Applications and Infrastructure
Vol. II: Functions, Operational Principles and Design
DAVID R. SLOGGETT, Marcol Group Ltd, Woking, Surrey
DIGITAL CARTOGRAPHY FROM SPACE
JONATHAN M. WILLIAMS, SERCO Limited, Sunbury on Thames

SERIES IN SPACE ENGINEERING

SATELLITE OPERATIONS: Systems Approach to Design and Control
JOHN T. GARNER and MALCOLM JONES, ESTEC, The Netherlands

SERIES IN SPACE LIFE SCIENCES

MARS AND THE DEVELOPMENT OF LIFE
ANDERS HANSSON, Editor-in-Chief, *Nanobiology*, Department of Medical Electronics,
St Bartholomew's Hospital, London, UK

SERIES IN REMOTE SENSING

APPLICATIONS OF WEATHER RADAR SYSTEMS:
A Guide to Uses of Radar Data in Meteorology and Hydrology
CHRISTOPHER G. COLLIER, Assistant Director (Nowcasting & Satellite Applications),
Meteorological Office, Bracknell
RACE INTO SPACE: The Soviet Space Programme
BRIAN HARVEY
SPACE MICROGRAVITY: Investigations in Earth-Orbiting Laboratories
ROYSTON G. W. HATHAWAY, Hayling Island, Hampshire
**REMOTE SENSING AND GEOGRAPHIC INFORMATION SYSTEMS: Geological Mapping Mineral
Exploration and Mining**
CHRISTOPHER A. LEGG, United Kingdom Overseas Development Administration, Forest
and Land Use Mapping Project, Forest Department, Colombo, Sri Lanka
INTERPLANETARY MISSION PLANNING AND SEQUENCING
D. M. WOLFF and W. I. McLAUGHLIN, Jet Propulsion Laboratory, Pasadena, California

REMOTE SENSING AND GEOGRAPHIC INFORMATION SYSTEMS:

Geological mapping, mineral exploration and mining

CHRISTOPHER LEGG
United Kingdom Overseas Development Administration
Forest and Land Use Mapping Project
Forest Department
Colombo, Sri Lanka

ELLIS HORWOOD
NEW YORK   LONDON   TORONTO   SYDNEY   TOKYO   SINGAPORE

First published in 1992 by
ELLIS HORWOOD LIMITED
Market Cross House, Cooper Street,
Chichester, West Sussex, PO19 1EB, England

A division of
Simon & Schuster International Group
A Paramount Communications Company

Printed and bound in Great Britain
by Redwood Press, Melksham

British Library Cataloguing in Publication Data

A Catalogue Record for this book is available from the British Library

ISBN 0–13–772336–9

Library of Congress Cataloging-in-Publication Data

Available from the publisher

Table Of Contents

Introduction

Introduction

Remote sensing has, over the past fifteen years or so, found widespread acceptance as a standard technique, on a par almost with geophysics and geochemistry, in many mineral exploration programmes, particularly in remote and less well mapped areas of the world. This is especially true in arid and semi-arid areas where vegetation cover does not obscure the rocks and the soils derived from them. The maturity of this use of remote sensing is indicated by the fact that special airborne scanners and even satellites are being developed to meet the needs of the explorationist. While some uses, for example in logistical planning, remain fairly constant, improvements in sensor design, in spatial and spectral resolution, and in the power and cheapness of image processing systems rapidly render any textbook on the subject at least partially out of date. There are many excellent texts on remote sensing applications in mineral exploration, but it was felt that the time had come for a new book which would incorporate the advances made in remote sensing in the second half of the 1980's. The possibilities for more extensive use of remote sensing in other aspects of the mineral industry are probably less broadly known and accepted. A series of studies at the United Kingdom National Remote Sensing Centre between 1985 and 1990 demonstrated the possibilities of remote sensing applied to environmental aspects of the minerals industry, in providing visual and quantitative information on land use and environmental impact at the planning stage, monitoring environmentally sensitive activities during mining, and then monitoring and assessing the success of post-mining restoration. The combination of growing environmental awareness, both within and outside the minerals industry, and increasing cost-consciousness in the industry could make remote sensing very much an idea whose time has come. The second main aim of this book is thus to bring the possibilities of remote sensing for cost-effective solutions to a range of tasks other than direct mineral exploration to the attention of those working in the industry. I hope that this book will both stimulate and encourage explorationists, planners, mining engineers and conservationists, and also act as an introductory guide to the principles and techniques involved. The book assumes a basic knowledge of the mining and quarrying industries, but does not require any previous acquaintance with remote sensing. I dedicate it to all our predecessors who battled, sometimes against formidable natural obstacles, to discover, mine and transport the minerals on which our society was built and on which it now depends, and who achieved all this without the considerable benefits which remote sensing can now bring to the industry.

The writing of this book could never have been contemplated but for the support and encouragement that I received from the manager and staff of the National

Remote Sensing Centre in Farnborough. I spent five and a half extremely happy and stimulating years at the NRSC, and had a unique opportunity to experiment with applications of a broad range of different types of imagery covering parts of many continents, and to meet users of remote sensing data from around the world. Mike Hammond, the NRSC manager, paid his scientific staff the compliment of allowing them great freedom of work towards clearly defined goals, in the belief that this would lead to the most innovative results. His confidence was justified, and a happy group of staff made major advances in a whole range of applications of remote sensing. My own interests turned from mineral exploration to the environmental aspects of mining, and the use of remote sensing in real time to monitor natural disasters, while others on the team, notably Tony Harding and Billy Loughlin, continued work on mineral exploration applications. The work of geographers, agriculturalists and oceanographers on the NRSC Applications Team, notably Nick Jewell, Geoffrey Griffiths, Jennifer Smith and Jan van Smirren, taught me much about non-geological applications of remote sensing, and many of the ideas in this book are derived from their work. The Geological Applications Working Group, later transformed into the Remote Sensing special interest group of the Geological Society of London, also provided me with the opportunity for much stimulating discussion during this period, chaired in succession by Kathy Bowden and Geoff Lawrence. My own full-time involvement in remote sensing results from contacts with, and later employment by, Hunting Geology and Geophysics, and the enthusiasm of Peter Martin Kaye, Eric Peters and Mike Barr was a great encouragement and support. Image processing is an essential part of satellite remote sensing, and I am grateful for all the assistance provided by George Kearn of the NRSC and Tony Ayles and Gill Garrard of Hunting Technical Services. Finally, this book would never have been completed without the sympathetic support of my wife Lorraine, who continued to believe that I could find the time and will-power to finish the text when I was ready to give up.

1
Principles Of Remote Sensing

1.1 DEFINITIONS

For the benefit of those readers, and I hope that they are the majority, who are not already experienced practitioners of remote sensing, I must start with a few definitions. Many other new features and concepts will be introduced as necessary in the text, and there is an appendix of most of the common acronyms at the end of the book. Remote sensing, like many other new sciences which have developed during the computer age, is burdened with a regrettably large number of unmemorable acronyms. There are even a growing number of second-level acronyms, or acronyms within acronyms, like FIFE, which stands for the First ISLSCP Field Experiment, where ISLSCP is the International Spaceborne Large Scale Climatology Project. I try to avoid too many acronyms and too much jargon, but these things are catching, and I apologise in advance for the few excesses and deviations which may have crept in.

It is first necessary to define remote sensing. This is both easy and difficult. Literally, remote sensing means the study of objects from afar without actually touching them, but this definition is so broad as to be almost useless. It would include microscopy, geophysics, astronomy, portrait photography and even looking at yourself in a mirror. The situation is no clearer in France and Germany, where the terms "Teledetection" and "Fernerkundung" are invented words with as many possibilities for ambiguity as the English. The use of the term is conventionally restricted to the acquisition of information, usually in image form, about the surface of the land masses and oceans, and the atmosphere above it, by airborne or spaceborne sensors. It is not restricted to passive sensors, receiving reflected or emitted electromagnetic radiation, since remote sensing includes airborne and spaceborne radar and lidar systems. The dividing line between remote sensing and airborne geophysics is far from clear, since there are passive geophysical techniques (gamma-ray spectrometry, magnetometry) and active techniques (airborne electromagnetic) which are not in principle very different from activities generally accepted as belonging to the family of remote sensing disciplines. What is generally accepted as belonging to remote sensing will become more clear as the reader continues with this book, while what is generally regarded as belonging to other disciplines will become apparent only by its absence.

GIS's (an acronym already!) are a subject on which much is written, to the extent that a curious kind of mystique surrounds them. Companies and government departments convince themselves and others that they are taking a major step forwards into a new technological age when they decide to invest in a GIS. In actual fact the minerals industry has been using GIS's for hundreds or even thousands of years. A standard topographical map is a GIS. It is a collection of georeferenced data sets, presented in such a manner as to illustrate their mutual relationships. A conventional 1:50,000 scale topographical map combines information about transport networks, land use, waterways, settlement patterns and elevation in a colour-coded fashion that displays a maximum of generalised information in a simple low-cost format. The big difference between a conventional paper map and the types of GIS that are the subject of so many conferences, journals and marketing drives at present is flexibility. The paper map presents a carefully selected and generalised set of data at a fixed scale and in a fixed format. A computerised, or more correctly, computer-hosted, GIS allows the production of specialised maps including any selected features from amongst a potentially very large set of attributes at a scale appropriate to each specific problem. It also allows the inter-relationships of numerous very large data sets to be studied and quantified. The power of modern computers, coupled to advanced display and hard-copy devices, allows rapid study of huge and disparate data sets in a way that was never possible with simple paper maps. Another "buzzword" can be introduced here. This is synergy, which is the concept of the whole being greater than the sum of the parts. The combination of many different types of map and image, for example geology, geophysics and geochemistry in a mineral exploration project, may reveal relationships and trends that were never apparent when examining either data set in isolation. There is thus nothing fundamentally new about the concept of a GIS, it is just the scale and speed of data handling which has changed.

Most computer-hosted GIS's permit the manipulation, display and analysis of geo-referenced data sets in vector or image (raster) form, and will also support a relational database of tabular information relating to areas and points on the maps. There will normally be a range of possibilities for data input, from tape drives for input of map and image information already in digital format, through scanners and video cameras for automated conversion of analogue maps or pictures to digital (usually raster) format, to digitising tablets for manual digitisation of maps. Output will be to a high-resolution colour monitor, and usually to printers and plotters, the latter sometimes being of very high quality for production of printing plates directly.

The next important definition is the image and associated nomenclature. A digital image is not the same as a photograph, and only becomes a picture when converted from digital to analogue form on a display screen or in a photographic print. The digital image is made up of thousands (usually actually millions in the case of satellite imagery) of discrete picture elements or pixels, arranged in regular rows and columns, each one of which has a digital number which relates to the average reflectance of the surface being imaged over the area covered on the ground surface by the picture element. This area is normally known as the resolution, or ground resolution, of the sensor, but is often mistakenly referred to as the instantaneous field of view (IFOV). The field of view should actually be the angular field of view of the sensor, with the ground resolution being dependent on the distance of the sensor from the surface. A digital image is acquired, imaged or scanned, but not "taken" as in a snapshot. A multi-band, or multispectral image is actually made up of a series of digital images, one for each waveband imaged, which can be considered as being stacked one on top of each other. For each pixel there will be a digital value for each wavelength imaged, the range of digital values in each pixel depending on the design of the sensor system. The electromagnetic radiation from each ground resolution element at each wavelength results in voltage changes in photoelectric detectors in the sensor. These analogue voltages are converted to digital numbers within the sensor system, and transmitted from the satellite to earth or recorded on magnetic tape. The raw digital numbers from the spacecraft are often further modified at the receiving station, but the most important factor governing the dynamic range of the data is the quantisation level in the sensor. This is the number of digital bits to which the original analogue signal is converted. In most remote sensing scanners, eight-bit quantisation is used, giving a possible dynamic range of between zero and 255 steps for each wavelength and pixel. Some meteorological satellites use ten-bit quantisation in order to achieve greater accuracy in temperature measurements, while the Japanese MOS-1 satellite, and its successor JERS-1, use six-bit quantisation, with a maximum dynamic range of zero to sixty-three. This reduces the amount of data to be transmitted or stored, but can seriously reduce the information content as well.

1.2 THE ELECTROMAGNETIC SPECTRUM

Electromagnetic radiation includes a very wide range of energy, from X-rays through visible light to radio waves. One thing that all types have in common is that they can be transmitted through a vacuum, and do not require movement of the molecules of the material through which they travel in the way that sound waves do. Only a portion of the huge electromagnetic spectrum is actually used for remote sensing, and is illustrated in Figure 1.1. The radiation is classified according to its wavelength, from 0.3 microns (a micron is one thousandth of a millimetre) in the ultraviolet region up to about one metre, the longest wavelength used for microwave remote sensing. Many specialist workers with microwave energy classify the radiation by its frequency rather than wavelength, and wavebands used for microwave remote sensing are in the gigahertz region (thousands of millions of cycles per second).

The upper and lower limits of the spectrum used for remote sensing are constrained by the physical interactions between the radiation and the materials which make up the Earth's surface. Since remote sensing is concerned with studies of the surface, and immediate sub-surface in a few cases, it is essential that radiation, either from natural sources such as the sun or from an artificial source such as an active radar system, be reflected, emitted or scattered back to the sensor in order to make an observation. When the wavelength of the radiation diminishes to the same order of magnitude as the spacings between the molecules in surface materials, the radiation is not reflected off the surface but penetrates it. It may then be diffracted or scattered, but only a small proportion of the incident radiation will find its way back to the sensor. This sets the lower limit of radiation usable for remote sensing to the ultraviolet region, although practical considerations discussed below actually set the practical limit for most applications at the lower end of the visible light region. The upper limit of the useful portion of the spectrum is set by the necessity to have reasonably detailed imagery of the surface of the Earth. The spatial resolution of imagery generated from electromagnetic radiation is restricted by the wavelength of that radiation, and effective resolution is probably at least ten times as large as the wavelength used. It would be possible to generate imagery at conventional radio broadcast wavelengths, but the resolution would be too coarse to be of any practical use.

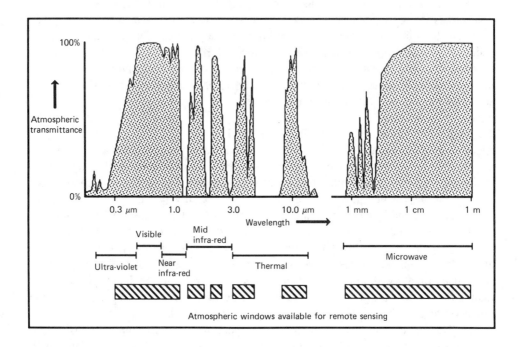

Figure 1.1. The portions of the electromagnetic spectrum used for remote sensing, showing main subdivisions and atmospheric windows.

Having set the upper and lower limits, it is now necessary to examine the subdivisions within the electromagnetic spectrum, and the special features of each of these subdivisions. The spectrum is divided into six main sections on practical grounds, influenced strongly by the absorption of large sections of the spectrum by water vapour in the earth's atmosphere. The usable portions cover so-called "atmospheric windows", wavelengths at which the atmosphere is essentially transparent. These include visible light, near infrared, mid-infrared, thermal infrared and microwave, as illustrated in Figure 1.1.

1.2.1 Ultraviolet

The ultraviolet region of the spectrum could be of great interest to geologists, since many minerals show characteristic fluorescence at these wavelengths, a feature which is used for mineral identification and sometimes in ground prospecting. The problem is that atmospheric absorption is very strong at this wavelength, and that little ultraviolet radiation from the sun actually reaches the earth's surface. This is a problem for remote sensing, but an advantage for most living things, which find excessive ultraviolet radiation toxic. Active ultraviolet remote sensing, using ultraviolet lasers on aircraft or helicopter platforms, has been tested for mineral exploration purposes, particularly for the detection of scheelite associated with gold mineralisation, and has also been used for monitoring oil slicks. The fluorescence of oil films on the surface of the sea can assist in the identification of the type and source of the oil. There are potential legal problems in the use of airborne ultraviolet laser systems, since the possibility, however remote, of such a laser shining directly into the eye of a human observer would be difficult to eliminate. The fluorescence induced by ultraviolet radiation is mainly in the visible portion of the spectrum, less subject to atmospheric absorption, and so it might be technically feasible to use a powerful satellite-mounted ultraviolet laser for remote sensing.

1.2.2 Visible Wavelengths

One of the largest atmospheric windows covers the visible wavelengths, those that are perceived by our unaided eyes. This appears a very fortunate coincidence until one realises that the existence of this window is precisely why animal eyes have evolved special sensitivity at these wavelengths. If there was strong atmospheric absorption in the region that we now term the "visible" portion of the spectrum, we might instead "see" in the mid-infrared or even in the thermal portion of the spectrum, and call these the "visible wavelengths".

Visible wavelengths are not the most useful for geological purposes. As shown in Figure 1.2, rocks, minerals and soils do not show distinctive spectral differences in the visible portion, just generalised differences in total brightness or albedo. Visible light is also more strongly dispersed by atmospheric haze, dust and pollutants than is radiation at infrared wavelengths. This results in a loss of contrast in visible wavelength imagery, low digital values being increased and high values decreased by atmospheric scattering

Despite these limitations, visible wavelength remote sensing is important in many fields. These are the wavelengths at which most conventional camera systems operate, and the comparative ease with which radiation at these wavelengths can be focussed and detected meant that most of the early spaceborne scanners operated in this region. Three of the four bands of the Landsat MultiSpectral Scanner, for a decade effectively the only spaceborne system available to civilian users, are at visible wavelengths, as are two bands of the French SPOT system. The comparison of radiance values in the visible and near infrared is essential to many standard measurements of vegetation vigour, and useful information on the relative iron contents of soils and weathered rocks can be derived from reflectance in the red portion of the spectrum.

The fact that visible light penetrates water to an extent dependent on the water purity, and is scattered back by suspended material, whereas radiation at infrared wavelengths is almost totally absorbed by water, makes visible wavelength remote sensing essential for studies of water quality, pollution and coastal bathymetry.

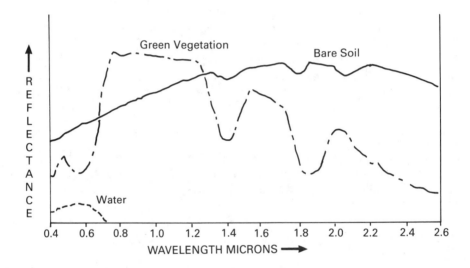

Figure 1.2. Spectral reflectance profiles for some representative natural surfaces.

1.2.3 Near Infrared
The near infrared portion of the spectrum is essentially a continuation of the main visible wavelength atmospheric window, and radiation at these wavelengths behaves in a very similar way to visible light. It can be focussed by fairly standard optical systems, and recorded by photographic emulsions, unlike the longer wavelengths in the mid-infrared. The total amount of electromagnetic energy available at a sensor is greater in this region that in the visible portion, and absorption and scattering by atmospheric pollutants is less. This results in a higher dynamic range for most sensors in the near infrared, and thus a higher signal to noise ratio. Imagery is usually sharp with good contrast, and is thus valuable for topographical mapping purposes. The strictly geological information content of this waveband, in terms of lithological or mineralogical discrimination, is not as high as at mid-infrared wavelengths, although

near infrared imagery is commonly used for structural studies. Ratios of radiances in the near infrared to radiances at visible wavelengths are very important in vegetation studies, and the almost total absorption of near infrared radiation by water, compared with the scattering of visible light, allows the use of this band for accurate discrimination of water areas in hydrological and related studies.

1.2.4 Mid-Infrared

This spectral region actually includes two "atmospheric windows", one centred at about 1.5 microns and the other at 2.2 microns, each with its own particular characteristics. The relatively low levels of radiation available at these wavelengths, the fact that this radiation is absorbed by all but the most exotic glasses used in optical systems, and the requirement for cooling of detectors operating at these wavelengths in order to maintain acceptable signal to noise ratios have restricted the use of this portion of the spectrum in remote sensing. This is a most important region for geological and vegetation studies, but the Landsat Thematic Mapper was the only operational spaceborne sensor providing mid-infrared imagery during the 1980s. The advent of specialised geological scanners operating in the 2.2 micron region on the Japanese JERS-1 satellite, and the inclusion of a 1.5 micron sensor on later versions of the French SPOT satellite, as well as coarse resolution 1.5 micron sensors on the ERS-1 ATSR and improved versions of the NOAA AVHRR should significantly improve the availability of data in this important region.

The 1.5 micron window is especially valuable for vegetation studies, permitting, for example, discrimination between broadleaved and coniferous woodland, and is also important in snow and ice studies. The 2.2 micron window, as indicated in Figure 1.2, is a region where many mineral species, particularly the phyllosilicates such as clay minerals, show characteristic absorption peaks. The inclusion of band 7 (2.2 microns) on the TM was mainly in response to demand from geological users, and has proved of great value in lithological mapping and in the discrimination of clay-rich alteration zones associated with mineral deposits, as described in Chapter 5 of this book. Airborne scanners, in particular the Australian Geoscan, have multiple bands within this window to distinguish different mineral species, and have proved of great value in mineral exploration. The sensors on JERS-1 are designed to provide similar information on a global basis.

1.2.5 Thermal Infrared

Reference to Figure 1.1 indicates that this region also includes two distinct atmospheric windows, separated by a wide atmospheric absorption region, mainly as a result of atmospheric water vapour. These windows are covered by a variety of sensors on operational remote sensing systems, as indicated in Figure 1.3. Remote sensing in this region is constrained by similar problems as in the mid-infrared, except that energy levels are generally higher in the 10 micron area. Conventional optics cannot be used, and detectors must be cooled to ensure acceptable noise levels. Development of spaceborne thermal infrared sensors has been driven by the meteorological community, who required accurate measurements of sea surface and cloud top temperatures, and the only operational spaceborne multispectral thermal scanners are the AVHRR on the NOAA satellites and the ATSR on ERS-1, each with one channel in the 3.5 micron window and two in the 10 micron region. These sensors are all carefully calibrated, and can provide very accurate temperature measurements of homogeneous surfaces. The single thermal band on Landsat TM is not so well calibrated, and generally gives only qualitative indications of temperature.

The direct measurement of temperature is rarely of interest in strictly geological studies, although the possibility of monitoring rangeland and forest fires (for which the 3.5 micron region is especially useful) and of mapping thermal effluent in rivers or lakes could be important environmentally. Geological interest depends more on differences in thermal properties between rock and soil units. Differences in specific heat, due particularly to variations in moisture content, can allow the mapping of concealed fault zones by use of pre-dawn thermal imagery, although existing satellite systems, with orbits optimised for land cover studies, do not acquire detailed imagery at this time. Differences in thermal emissivities can be of great value in lithological discrimination, but their detection requires a multispectral capability in the 10 micron window. This is presently available only in the airborne TIMS system, which has generated impressive imagery in some geological campaigns.

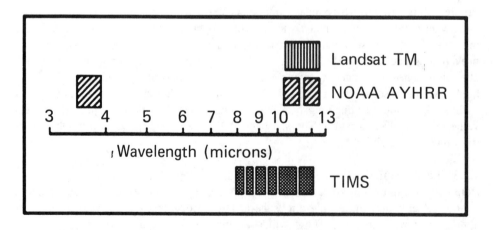

Figure 1.3. Wavebands of thermal infrared sensors.

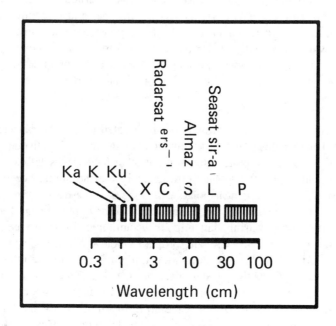

Figure 1.4. Nomenclature of microwave bands used in remote sensing,
with examples of spaceborne systems.

1.2.6 Microwave

The microwave portion of the spectrum can be rather arbitrarily divided into a series of bands, as indicated in Figure 1.4. These subdivisions are historical, and derive from aviation military and civilian radar systems, rather than from remote sensing of the type that we consider here. In general, the short wavelengths are most strongly absorbed by natural materials, especially water, while the longer wavelengths penetrate further into soils and overburden, especially if they are dry. The amount of energy scattered back to the sensor depends on numerous factors, but especially on the dielectric constant of the material, its surface roughness as a function of the wavelength being used, and the relative inclination of the surface with respect to the radar beam. The last factor often dominates, so that topography usually has a critical influence on the appearance of radar imagery. The dielectric constant will be a function of lithology, but also of moisture content, while surface roughness will sometimes be characteristic of certain lithologies. The lithological discrimination possibilities of a single wavelength of microwave imagery are probably rather limited, since the interactions of different factors are difficult to model. Topographical correction using digital elevation models might assist, but this is extremely computer intensive. Until the advent of multispectral microwave systems in the late 1990's, the main use of microwave remote sensing in the minerals industry is likely to be as an aid to structural interpretation. The sensitivity of backscatter to topographical differences emphasises structural features, although preferentially in directions at right angles to the radar beam, and skilled photo-interpreters can extract much useful information from radar imagery. In many parts of the world, especially the equatorial rainforests, cloud cover prevents the acquisition of imagery at optical wavelengths. Microwave remote sensing is then the only source of data, and must be used as a sensor of last resort.

1.3 ORBITS

There are three main classes of orbit into which satellites can be placed. Equatorial orbits, as the name suggests, are those in which the satellite orbits the earth in or near the plane of the equator. In polar orbits, on the other hand, the satellite orbits over or near the North and South Poles. The third category are the highly eccentric Molniya orbits, usually but not always near-polar, where the satellite travels far out into space on one side of the earth and then passes very close to the opposite side. Molniya orbits are not used in remote sensing, but only for communications purposes in extreme northern latitudes. Although there are proposals for a remote sensing satellite in equatorial low earth orbit, the only current equatorial orbit used for remote sensing is a very special case known as the geostationary orbit. It is a fundamental law of physics that the orbital period, the time taken for a satellite to complete one complete rotation of the earth, increases with the radius of the orbit. In simple terms, the further out into space that a satellite is, the longer it takes to complete an orbit. There is thus a specific orbital altitude at which the orbital period is equal to the period of rotation of the earth, one sidereal day. As pointed out more than forty years ago by Arthur C. Clarke, a satellite in such an orbit will appear to remain fixed over the same point on the equator, hence the name geostationary orbit. This orbit is widely used for communications satellites, since it permits continuous communication between two points from which the satellite is in view, without the need for costly and complex movable antennas. This orbit is also used for remote sensing, although since the satellites are far out in space (35,000 km), all current systems provide only a rather crude picture of the earth's surface. Geostationary orbits are mainly used for meteorological remote sensing, where great surface detail is not required, but very frequent images are essential.

Most remote sensing of the earth's surface is carried out from polar orbits. The reasons for this are two-fold. Firstly, the earth rotates beneath the satellite track as it travels from pole to pole. The satellite orbit is essentially fixed in space relative to the centre of the earth, but the earth itself rotates in a plane approximately at right angles to the plane of the orbit. This permits a single satellite to eventually cover the entire surface of the earth in successive orbits. The second reason is that by making the orbit slightly offset from the poles, and oblique to lines of longitude, the local sun time of each point along the orbital track will be the same. During the time that it takes for the satellite to travel from overhead one pole to overhead the other, the earth will have rotated slightly, and the orbit can be planned to follow a line of equal solar time on the earth's surface. In practice, most earth observation satellites are in sun-synchronous orbits and acquire images at between 0930 and 1030 local time. This is regarded as being the time of least cloud cover in most climates, and also provides oblique illumination which highlights relief in satellite imagery. In multi-date remote sensing studies, it is an advantage to have all images acquired at the same local time, since this minimises illumination differences between them, although seasonal differences in solar elevation cannot of course be eliminated by this means.

As stated earlier, the orbital period is governed by the height of the orbit. Most earth observation satellites are at altitudes of between 700 and 900 kilometres, giving orbital periods of between 98 and 103 minutes. This altitude range is selected as a workable compromise between the requirement, in some applications, for the best possible spatial resolution, and the need for the satellite to remain operational for as long as possible. Lower orbits at about 300 km, such as those used by the American Space Shuttle, would permit the acquisition of more detailed imagery, and even lower orbits are used for specialised military remote sensing satellites. In low orbits, however, the effects of drag from the extremely thin atmosphere still present at those levels become serious. The satellite is slowed by this drag, and drops gradually to lower orbits. If a constant orbit is to be maintained, which is essential for most remote sensing applications, fuel must be used to speed the satellite and thus lift it to a higher orbit. The amount of fuel carried on a satellite is strictly limited both by storage capacity and by the fact that all this additional weight has to be lifted into orbit when the satellite is launched. Some large Soviet remote sensing satellites in low earth orbit carried as much as four tonnes of fuel at the time of launch, and even then had a life of only a few years.

1.4 SATELLITES

Earth observation satellites are extremely complex pieces of engineering, and vary greatly in design. There are however features in common amongst all these satellites, and all include the same basic elements. All remote sensing satellites have a power supply, an instrument or set of instruments for observing the earth, a means of transmitting the information from these sensors down to earth, equipment for receiving operating instructions from ground stations, and for controlling the attitude and position of the satellite. All these components are housed in or mounted on some kind of platform, which will include the means for communicating between the different components, distributing power to them, and controlling temperature. In some cases a standard design of platform or "bus" , complete with power supply, attitude control, temperature control and other "housekeeping" functions, is used for many different missions, being fitted with different instruments as required. This is true, for example, of the NOAA series of satellites operated by the American National Oceanographic and Atmospheric Administration, and of some series of Soviet satellites. A well-proven design minimises costs and the chances of in-flight failure.

In almost all cases, the power supply for earth observation satellites is generated by panels of solid-state solar cells. The solar panels must be steerable so as to continue facing the sun as the satellite orbits the earth. Since no power will be generated when the satellite is orbiting the dark side of the earth, batteries must be provided to maintain power for almost 50% of each orbit. The power requirements of passive optical sensors are not large, but active microwave sensors require much greater power. In many microwave satellites, the active radar system only operates for a small proportion of each orbit, the remainder of the orbit being used to charge the batteries. The exceptions in terms of power supply were some Soviet military radar systems, where the power requirements were too large to be met by solar energy, and where small nuclear reactors were used instead. International objections to nuclear reactors in space, especially in the low earth orbit used by military satellites, have largely deterred the use of such power plants, even though they might have considerable advantages over solar energy where power requirements are large and great flexibility of operation necessary. A nuclear power supply is also used on some deep-space probes, which are designed to travel so far from the sun that little energy would be available from this source.

Stabilisation of an earth observation satellite is most important. The instruments must point in the correct direction, and the orbit must be maintained extremely precisely if imagery is going to be acquired repeatedly from identical areas on the earth. All remote sensing satellites in polar orbits are what is known as "three axis stabilised", which means that the pitch, yaw and roll motions of the spacecraft are all precisely controlled. This control is usually achieved by a combination of techniques. Relatively heavy wheels, similar in operation to a flywheel, may be built into the satellite, and rotated in the opposite direction to any unwanted movement so as to react against it. Thrusters, essentially miniature rocket motors, are also used to counteract unwanted rotation or other movement, although these consume fuel and must be used sparingly if the satellite is to have a usefully long life. Thrusters are also used to control the speed, and thus the altitude of the satellite, in order to maintain a constant orbital period and thus a regular repeat cycle. A series of sensors on the satellite, usually one or more detecting the location of the sun, others detecting the earth, and others based on inertial platforms which monitor rotations, provide the information necessary for precise attitude control. In many cases the orbit of the satellite is also tracked precisely from earth, and instructions transmitted to the satellite for orbit corrections.

Temperature control is very important in satellites. They are exposed to extreme changes in temperature as they pass from sunlight into shadow, and most of the instruments and electronics on board must be maintained at constant temperatures in order to function correctly. As will be discussed in a later section, most remote sensing scanners have detectors which must be maintained at very low temperatures. Temperature control is achieved firstly by insulating the interior of the spacecraft from solar heating. The outside of the satellite is covered with highly reflective material, often gold foil, to reflect solar radiation and thus minimise absorption of heat. Heat generated within the satellite by electronic components and motors is conducted out to the side of the spacecraft which is, in sun-synchronous orbits, never illuminated by the sun, and is then radiated out into space.

Some satellites carry complex recording systems, usually based on magnetic tape, to record imagery acquired by scanners for later transmission back to earth. All satellites require some kind of onboard memory to store commands and operating procedures determined before launch or transmitted from earth. There must also be one or more computers to coordinate and monitor all the operations of the spacecraft.

All the components of a satellite must be constructed to extremely rigorous specifications. Everything must function perfectly in a very hostile environment for many years without the possibility of maintenance or repair. Some duplication of systems is usually possible to provide back-up in case of breakdown, but limitations of weight and space do not permit complete redundancy. There have been a few examples of in-orbit repair of satellites using the American Space Shuttle, but this is not normally feasible since the Shuttle is in low earth orbit and most remote sensing satellites are in higher orbits. For reasons of durability, the latest "state of the art" electronics are often not used in satellites since they have not been tested for long enough periods in harsh environments. Most satellite electronics are in fact some years behind advanced earth-bound systems. This situation is also partly a result of the time required to design, build and test a new satellite before launch. As an example, the European Space Agency ERS-1 satellite, launched in July 1991, was mainly constructed before 1986, and most of the electronics date from the early 1980s.

1.5 DATA RECEPTION

All digital imaging systems on satellites must transmit their data down to ground stations if the data is to be of any use. Camera systems, notably on Soviet satellites, must return exposed photographic film to Earth, a complex technology. Digital data is transmitted to Earth at radio wavelengths, some satellites transmitting continuously, while others do so only on command from the ground or after recording part of an orbit of data.

The complexity and cost of receiving stations depends mainly on the amount of data that is to be received, and this in turn depends on the spatial resolution and number of wavebands of the sensor. The satellite orbit also has an effect on the cost of data reception, since satellites in polar orbit normally require steerable antennas which track the satellite as it moves in relation to the ground station. Geostationary satellites do not require steerable antennas, and data can be received with much lower cost fixed dishes. The data rate for Meteosat imagery is only 0.166 megabits per second, allowing reception with a low-cost and simple receiver. For this reason, most users of Meteosat data have their own receiving stations and do not rely on central ground stations or archives. The next increment in terms of data rates amongst earth observation satellites is the AVHRR sensor on the NOAA satellites. This transmits information back at a rate of about 4 megabits per second, significantly more than Meteosat, but still within the range of relatively simple receivers and the recording speed of a conventional hard-disk system. The drawback is that the NOAA satellites are in polar orbit, and hence the antenna must normally be steerable and precisely controlled for full horizon to horizon coverage. This steerable antenna is usually by far the most expensive single item in an AVHRR receiver. If all that is required is imagery covering the area around the observer, a simple fixed antenna will suffice, and low-cost systems for reception of AVHRR imagery are now being installed in many countries by data users. Satellites such as the Indian IRS-1 and Japanese MOS-1 transmit at total data rates of about 25 megabits per second, and this order of data transfer begins to require very specialised equipment. Antennas must be large to increase signal to noise ratios and thus minimise bit error rates, steering and programming of the antenna must be very precise, and high density data recorders must be used. Data rates increase still further with finer spatial resolution and more spectral bands, SPOT and Landsat TM having data rates of 50 and 85 megabits per second respectively. This tends to limit reception to national and regional facilities, and this mode of reception is actively encouraged by the satellite operators, who are partially dependent on data sales, and wish to maintain some degree of control over the reception and distribution of their data.

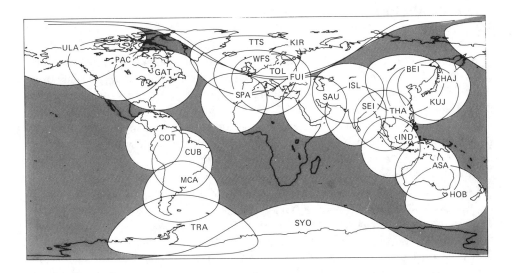

Figure 1.5. ERS-1 ground-stations, actual and proposed (courtesy ESA).

The distribution of ground stations for ERS-1 is shown in Figure 1.5. Many of these stations also receive TM and SPOT imagery, and there are also some specialist SPOT ground stations. The area of coverage of each station is indicated on the map, and it can be seen that there are large areas of the earth which are not within the reception range of any ground station. With SPOT this is not a serious problem, since the satellites carry on-board recorders and global coverage can thus be obtained through the main control and reception stations in Sweden and France. Currently operational Landsat satellites do not carry recorders, and global coverage is theoretically obtained through a network of geostationary data relay satellites. Unfortunately, the demands from other users, notably the Space Shuttle, for the relay satellites has resulted in a coverage gap over an area of Asia including much of India. Within this gap, imagery can only be obtained through ground stations in the area.

1.6 ARCHIVING AND DISTRIBUTION

Satellite image archives are often maintained by data users themselves, in the sense that they store a selected set of satellite imagery that has been purchased for a specific purpose. This is not the kind of archive that we are discussing here. This section is about archives maintained by satellite operators, distributors of satellite data and national data centres. The main reasons for them to maintain an archive are either that satellite data is their stock in trade, and imagery covering different areas at different dates must be stored so that it can be sold or otherwise supplied to customers, or else that data from a specific satellite must be archived for long-term research purposes. A digital data archive is not like a library, neither is it comparable to a bookshop. The data is in most cases for sale, and not for loan, but because of the nature of digital data it is not necessary to store multiple copies of every product waiting for a customer to purchase them. Copies can be generated from master image files when a customer requests them.

In most satellite data archives, the basic storage medium is magnetic tape. If the archive is connected to a receiving station, the raw imagery is usually stored in the high-density form in which it is received. These HDDT's are then converted to standard computer compatible tapes (CCT's) at customer request. Some processing of raw data is usually carried out during the conversion from HDDT to CCT. The digital values are re-scaled using standard gain and offset values to occupy a full eight bits per band, and a crude geometric correction (de-skewing) is usually carried out to compensate for earth rotation beneath the satellite during imaging.

Magnetic tapes are not necessarily the ideal medium either for storage or for distribution of image data. They have become the accepted standard for large volumes of data because of their relative ease of use, and now that most users have tape drives, this is the dominant medium for distribution. Magnetic tapes suffer from serious disadvantages, however. Firstly, they are not a stable storage medium. Tapes can be corrupted by storage in improperly regulated environments, where temperature and humidity are not precisely controlled, and where stray magnetic fields are present. Tapes need to be re-wound at regular intervals to prevent data from "printing through" from adjacent layers. Tapes also occupy a large space relative to other media, and are bulky to transport as well as to store. The rates of data writing and reading are higher with tapes than with some other media, but are still relatively slow. Another drawback is that a tape archive cannot be physically "on-line". The required tape must be manually selected from a storeroom and loaded onto a tape drive before the data can be inspected or used.

Optical disks have many advantages over magnetic tapes, but are still not in wide use as archive or distribution media. One reason for this is certainly a lack of standardisation in the disks themselves, the drives which read and write them, and the file structures. There are numerous formats available, and optical disks written on one system cannot often be read by another. Another problem may be that long-established archives are usually on magnetic tape, and the cost of transferring large volumes of data to optical disk may be intimidating. Optical disks have the advantage of extreme durability, they occupy a much smaller volume for the same amount of data stored than do magnetic tapes, and it is possible to have a fairly large archive "on-line" in the sense that disks can be loaded automatically in a "juke-box" system, and any image or portion of an image can be displayed on a terminal within a fairly short time. The write and read times of optical disks are slow compared with some magnetic tape systems, and other storage media are now coming into use which may have the desired combination of durability, small size and speed and flexibility of access.

A satellite data archive, like a library, must have a catalogue. For most users or purchasers of the data, a simple list of dates of acquisition and areas covered will not be sufficient. Cloud cover or haze may obscure the area of interest, which may only be a small portion of a whole image, and with some satellite systems the orbit is not precisely controlled, leading to large differences between the theoretical geographic coverage of a scene and the actual area covered. The data user needs to be able to view the image data, usually in a degraded form, in order to select an appropriate image for his needs. This can be achieved in a number of ways. The traditional approach has been to produce a reduced spatial resolution single-band monochrome photographic image from the incoming data at the time of reception from the satellite. Most Landsat receiving stations generate these so-called "quicklooks" which can either be viewed at the receiving station or through its appointed distributors, or sent to the potential customer at a small charge. Some receiving stations use video systems to record the single-band image, and a whole sequence of images can be viewed on a video screen to select suitable scenes. A "quicklook" image can also be stored in digital form, sometimes in a multispectral format, allowing customers to view imagery on special terminals, which can be remote from the actual archive. An optical disk or other on-line archive allows the possibility of viewing any part of an archived image at full resolution, and of carrying out simple enhancements on it prior to ordering.

Once the customer or other user has chosen the imagery required from an archive, then the data must be delivered so as to be available for use. The format of the image data, in terms of file structure, block-length and so on, must be such that users can read the imagery on their computer systems, and data distributors have developed a series of standard formats for satellite imagery, such that most image processing systems can read imagery from any major distributor. The dominant distribution medium is the magnetic tape, since tape formats are standardised and most organisations with a computer will have a tape drive. Very few data distributors supply data on optical disk, probably mainly because of the lack of standardisation. Some archives will supply data on tape cartridges, readable on lower-cost drives than conventional tape drives. The supply of satellite imagery on floppy disk is not a practical proposition, except for educational purposes. Floppy disks normally have a maximum storage capacity of 1.4 megabytes, and when it is realised that a single 7-band Landsat TM scene occupies 240 megabytes, the impracticability of this medium is obvious.

1.7 SENSOR SYSTEMS

The sensor systems used in remote sensing are of two very different types, with important subdivisions within the main types. The two main groups are the passive and active sensors, the former simply recording reflected or emitted electromagnetic radiation, while the latter produce their own illumination and record it's effects. Within the passive sensors are included photographic (analogue) systems and scanners (digital) while active systems include imaging radar and laser (lidar) systems. Scanners can themselves be divided into two main groups, the optomechanical scanners, where a rapidly moving mirror is used to scan the surface below in a pre-determined fashion, and the pushbroom scanner, which has no moving parts, but records radiation from the surface below by means of arrays of sensitive semi-conductors (usually charge-coupled devices or CCDs), the number of detectors in each array being equal to the number of pixels in each row of the image. Airborne and spaceborne scanners do not differ essentially in design.

A typical optomechanical scanner is the Landsat Thematic Mapper (TM). The basic layout of this system is illustrated in Figure 1.6, and is similar in principle to the MSS, and to the commonly used Daedelus airborne scanners. Electromagnetic radiation from the earth's surface is directed into the optical system of the instrument by a rapidly oscillating mirror. The radiation is then focussed by a Cassegrain-type reflecting telescope onto a group of solid-state detectors at the primary focus. These are uncooled detectors, sensitive to the three visible bands and the near infrared band of the TM. A portion of the radiation continues past the principal focus, and is re-focussed by means of an additional pair of mirrors onto another set of detectors at the secondary focal plane. These detectors operate in the mid-infrared and thermal portions of the spectrum, and need to be cooled to extremely low temperatures in order to be stable and sensitive. In the TM there are actually sixteen detectors for each wavelength, and adjacent pixels on sixteen successive scan lines are imaged simultaneously. The reasons for this must be sought in the principles of scanner design, and the complex trade-offs between spatial resolution and signal quality. Forward scanning, from line to line is achieved by the forward motion of the spacecraft, and in the simplest type of scanner, with only a single detector per waveband, the mirror must move sufficiently fast that it scans a whole line of pixels in the time taken to travel forward a distance of one pixel. The smaller the pixel, or the finer the spatial resolution, the faster the mirror will have to move. A corollary of this is that small pixel sizes will permit ever shorter integration times for measurement of the radiation from each pixel on the ground surface, and could result in increases in random noise within the system. The AVHRR scanner on the NOAA meteorological satellites has only a single detector per waveband, and the mirror actually rotates through a complete circle of 360 degrees, but acceptable integration times are achieved because of the very large pixel size of about one kilometre. At the spatial resolution of TM (30 metres), this would not be possible. Sufficient time for the measurement of the radiation from each ground pixel is achieved firstly by imaging sixteen lines simultaneously, and secondly by using an oscillating rather than a rotating mirror, with scanning in both forward and reverse directions. This results in an instrument of considerable mechanical complexity, and it is a tribute to the high manufacturing standards of the TM that it has operated continuously for many years in the harsh environment of space, and that the image quality is generally acknowledged to be excellent.

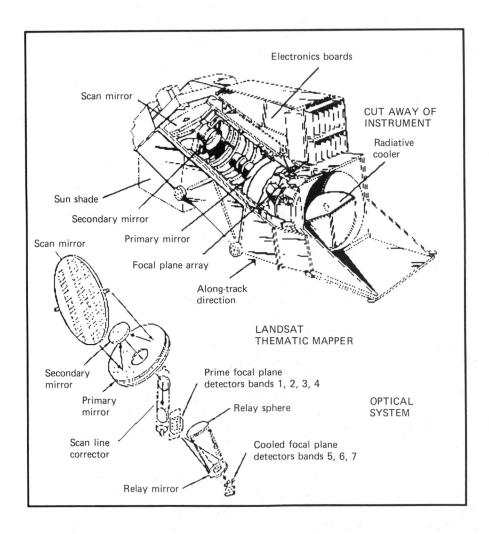

Figure 1.6. The Landsat TM, an advanced spaceborne optomechanical scanner.

Some of the problems of mechanical scanners are reduced in the pushbroom scanner, a typical example of which is illustrated schematically in Figure 1.7. The pushbroom scanner has no moving parts, but images the whole of a scan line at one time using a very large array of detectors, usually charge coupled devices or CCD's. The early use of pushbroom scanners was delayed by the problems of manufacturing large arrays of identical detectors, and the first commercially operational examples were used in the French SPOT satellite, launched in 1986. The Soviet MSU-E scanner, the MESSR on MOS-1 (Japan), the LISS scanners on the Indian IRS-1 and the Canadian Moniteq airborne scanner use essentially the same principles. Electromagnetic radiation from the ground is focussed by a lens or mirror system through a slit and onto a diffraction grating or similar device to separate it into the required spectral bands. Each wavelength is then focussed onto a separate array of CCD's, which are sampled electronically in sequence to achieve across-track scanning. This sampling must be completed in the time that the spacecraft moves forward one pixel, when the next scan line is imaged. Finer spatial resolutions can be achieved since there are no restrictions on mirror speed, and integration time for each pixel is the time taken for the spacecraft to move forward one pixel, rather than a very small fraction of that in an optomechanical system.

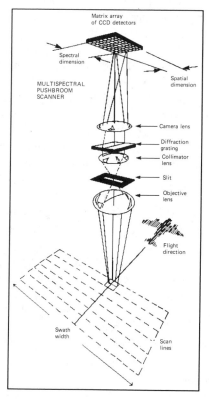

Figure 1.7. Schematic diagram of an airborne pushbroom scanner.

Both types of scanner, mechanical and pushbroom, have their own characteristic types of systematic noise, caused by small differences in sensitivity between different detectors in the arrays, whether they be arrays of seven, eight or sixteen detectors in the optomechanical systems, or many thousands in pushbroom devices. Differences in detector response result in different digital values in the final image. In the case of optomechanical scanners, these differences appear as systematically repeated across-track bands of slightly different average digital values, while in pushbroom systems the banding, which is not repetitive in character, is in the along-track direction. Methods of minimising the effects of banding are discussed in Chapter 2.

In both types of scanner, electromagnetic radiation reaching the detector produces a voltage which is proportional to the amount of incident radiation. This voltage is sampled at regular intervals by the electronics of the scanner, and converted from analogue voltage to a digital value. The total dynamic range of voltages for which the system is designed is subdivided into a pre-set number of digital steps, or quantisation levels. For the Landsat and Spot systems, eight-bit quantisation is used, permitting 256 grey levels for each waveband. The AVHRR scanner on the NOAA satellites has ten-bit quantisation in order to accommodate the broad dynamic range of thermal measurements, while the Japanese MOS-1 and JERS-1 systems use six-bit quantisation. In most spaceborne scanners the gain levels are pre-set before launch, so that the sensors cannot respond to changes in the average radiance of a scene. Since polar orbiting systems pass, on each orbit, from polar regions with relatively low solar elevation angles, through the tropics with high sun angles, and back to polar regions again, the detector gain must be set so as not to saturate even over equatorial regions. This results in the range of digital values at high latitudes being considerably less than the total dynamic range of the system, especially in winter. Some sensors have high gain modes for special operations. The dynamic range of radiances over water is normally very low, and the Japanese MOS-1 MESSR sensor has a high-gain mode specifically for oceanographic studies. In this mode the sensor would saturate over land. The main restriction on the use of large ranges of quantisation levels, for example ten or even sixteen bits, is the increased volume of largely redundant data that needs to be transmitted and archived. Much work is in progress on techniques for data compression in order to allow large dynamic ranges without the penalties of large data volumes.

Some observations in the microwave portion of the spectrum are also made with passive scanners, which are essentially very sensitive radio frequency receivers with highly directional antennae. Passive microwave sensors usually operate at a range of wavelengths, and scan the ground surface either by physical movement of the antenna or its waveguide or by electronic scanning of a phased array system. Despite the directionality of the antennae, passive microwave sensors operating from orbital altitudes cannot achieve fine spatial resolution at the earth's surface, and the effective pixel size is usually in excess of five kilometres.

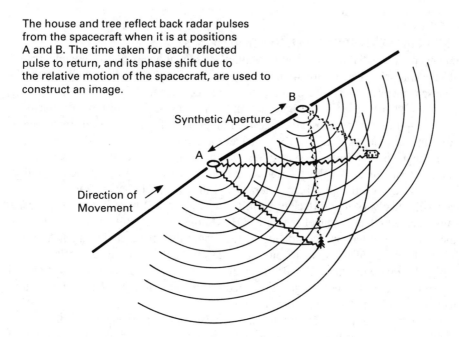

The house and tree reflect back radar pulses
from the spacecraft when it is at positions
A and B. The time taken for each reflected
pulse to return, and its phase shift due to
the relative motion of the spacecraft, are used to
construct an image.

Synthetic Aperture

Direction of
Movement

Figure 1.8. Schematic diagram showing the principle of SAR.

Most remote sensing in the microwave portion of the spectrum which is of
interest to the minerals industry is carried out with active sensors, or radar. For an
active microwave sensor to produce imagery of fine spatial resolution, the antenna size
must be a significant multiple of the wavelength of the microwave energy employed.
Real aperture radars, using large antennas, were feasible in airborne systems, and were
widely used for mapping, particularly in equatorial rainforest areas, during the 1960s
and 1970s. Such large antenna systems are not feasible for spacecraft, however, and
even for airborne work are very unwieldy. A more practical solution is to use the
forward motion of the satellite or aircraft to simulate a very long antenna orientated in
the direction of movement. Each point on the ground is imaged by a number of

successive microwave pulses from different positions as the antenna moves forward, creating the effect of a very long fixed antenna. The principle is illustrated in Figure 1.8. The problem with these synthetic aperture radar (SAR) systems is that processing of the raw SAR data to produce a digital image is extremely complex. The multiple returns from each point imaged have to be separated from the returns from all other points, a process which uses Doppler shifts in the frequency of returned pulses from objects approached and then passed by the sensor. Great increases in computing speed and power, and increasingly sophisticated software, now allow the generation of digital SAR images in something approaching real-time. When data from the first spaceborne SAR, on the Seasat satellite launched in 1978, was processed, each scene required up to two days of computer time.

1.8 SPATIAL AND SPECTRAL RESOLUTION

The concepts of spatial and spectral resolution are sometimes seriously misunderstood, as are the compromises that are almost always required in the design and operation of a remote sensing system. This section attempts to remove some of the misunderstanding, and to assist the new user in the choice of appropriate resolutions for the task in hand.

Spatial resolution is usually described in terms of the pixel size of the imagery, for example ten metres in SPOT PAN imagery and one kilometre in AVHRR. This is not an adequate definition of spatial resolution. The actual resolving power of the system, in terms of the smallest surface feature that can be seen, is a function both of the pixel size and of the optical characteristics of the feature. In near infrared imagery a stream will have very low reflectance whilst green grass will have high reflectance. A stream much less than the pixel size of SPOT XS imagery can still be detected because of the effect that it has on the radiance of the whole line of pixels in which it occurs, although the width of the stream cannot be measured unless the precise reflectance of water and grass are known, and the average reflectance can then be modelled. Conversely, features much larger than pixel size may not be detectable if their reflectance closely matches their surroundings.

Spectral resolution, strictly defined, means the narrowness of the spectral channels used in the sensor. A high spectral resolution system will have very narrow channels, imaging precisely defined portions of the spectrum. In practice, the term high spectral resolution is used for systems with numerous spectral channels. Landsat TM is considered to have better spectral resolution than SPOT, with seven channels against three. All currently operational spaceborne remote sensing systems image only small selected portions of the total electromagnetic spectrum. Systems which image a continuum of a portion of the spectrum, for example the visible, near infrared and mid infrared, are known as spectrometers, and may have as many as 256 narrow channels contiguous with each other. These instruments provide extremely valuable information on the spectral response of different surface materials, enabling optimal design of future remote sensing systems. Much of the information in the majority of channels is probably redundant for any given surface, and it is unlikely that future spaceborne systems will operationally image a continuum. Specific narrow channels providing critical information about surfaces of interest will probably be imaged selectively.

There is a trade-off between spatial resolution and acquisition frequency, and between spatial resolution and cost per unit area of imagery, as shown in Figures 1.9 and 1.10. Constraints in data handling and transmission rates, as well as in the optical and electronic design of sensors, limit the width of images in terms of the numbers of pixels per line. For commercial systems, the maximum is currently 6000 pixels per line in SPOT PAN imagery. If fine spatial resolution is required, this limits the width of the image, and thus the width of the earth's surface that is imaged during each orbit. The SPOT system, with its two parallel scanners, images about 120 kilometres width of the surface during each orbit. Landsat images 185 kilometres, while the AVHRR images more than 3000 kilometres, although at a greatly decreased spatial resolution. If the system is to achieve repetitive global coverage, which is a requirement of most operational remote sensing, then the repeat frequency is controlled by the swath width. The narrower the swath width, the longer the interval between repeat coverage. This obviously has implications for monitoring operations, as compared with simple mapping. If dynamic, rapidly changing features need to be monitored, then a high frequency of repeat coverage is required, whilst periodic mapping may require a much lower repeat frequency. Present systems only allow high temporal frequency imaging at coarse spatial resolutions, and the only way of obtaining fine spatial resolution imagery at high repeat frequencies is to have many more satellites of SPOT or Landsat type, operating on complementary orbits so as to give more frequent imaging.

The inverse relationship between spatial resolution and cost per unit area is a natural consequence of data handling costs, and also reflects the commercial imperatives of the satellite operators. The main implication for users is that spatial resolution should be matched to requirements. There is little point in buying imagery of the finest available spatial resolution if a regional overview of a whole country is required. The choice of appropriate imagery is discussed further in Chapter 4.

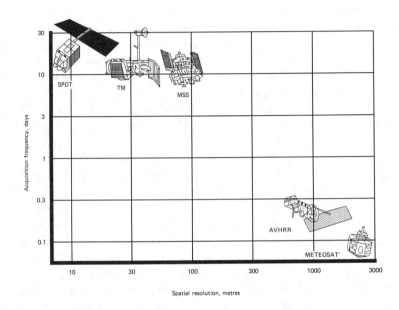

Figure 1.9. The relationship between spatial resolution and acquisition frequency
for some common spaceborne sensors.

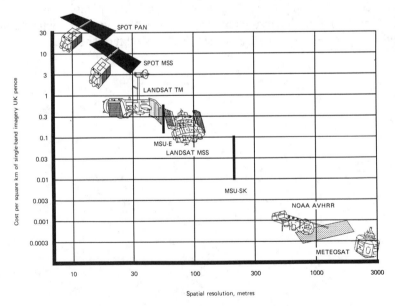

Figure 1.10. The relationship between spatial resolution and cost per unit area
for some common spaceborne sensors.

2
Image Processing

Image processing is a vital part of most remote sensing operations, even if the final end user is not conscious of this. All digital imagery must be processed in some way in order to be of use in the vast majority of applications. The image has to be displayed on a screen for visual interpretation, photographic hard copy has to be produced for the same purpose, or derived products such as maps and tabular information must be extracted. In many cases, more than just the production of a simple monochrome or colour image is required, and more complex processing is needed in order to produce a final product for the user. Image processing can be divided into two distinct stages, the first of which (pre-processing) is essential in almost all operations using digital imagery, and the second (image enhancement and information extraction) required in all the more sophisticated applications.

This section is not intended to be a "do-it-yourself" guide to image processing. There are many excellent textbooks on the subject, some of them listed at the end of this book. The intention here is to give a brief overview of the more important image processing operations in order that the reader has a better appreciation of the possibilities of and necessity for the various techniques. Even if readers never have to actually carry out their own image processing, they will be, hopefully, better informed when talking to specialists in this combination of science and art, as for example when negotiating with remote sensing contractors or assessing the recommendations of their in-house remote sensing unit.

2.1 PRE-PROCESSING
The first part of image processing is usually known as pre-processing, since it must precede most other image processing operations. The amount of pre-processing required will vary with the sensor type and the quality of the digital data, and also with the use to which the imagery is put.

2.1.1 Earth Rotation Correction
Polar orbiting satellites are usually in slightly eccentric orbits in order to attain repeat coverage of each part of the world at roughly the same local sun time, and since the earth is rotating beneath the satellite as it passes overhead, images will never be aligned in a correct north - south direction. Each line of the image should be slightly offset from the line before it to compensate for earth rotation, and an early stage in most pre-processing operations is an earth rotation correction which transforms a square or

rectangular image into a rhomboidal image, improving the relative positional accuracy of features in the image. The effect of this is illustrated in Plate 1, a Landsat MSS scene after earth rotation correction. The transformation of an originally square image into a rhomboid which more accurately reflects the path of the satellite over the earth's surface can be achieved by complex re-sampling of the pixels, although this would be computer-intensive. The correction is more usually undertaken by shifting every nth line, where n depends on the obliquity of the satellite orbit, along by one pixel. This does not modify the original pixel values, and has little visual effect on image features.

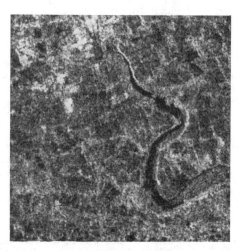

Figure 2.1. Random noise in remotely sensed imagery. The image on the left, from the Soviet MSU-SK sensor (courtesy Glavkosmos), was acquired at visible wavelengths, while that on the right, from Seasat (courtesy RAE), is a microwave image.

2.1.2 Noise Reduction

Most digital images contain noise. This may be random noise, generated in the many stages of electronics that the signals pass through between the original sensors on the spacecraft and the image processor used by the observer. Examples of random noise seen in imagery from optical and microwave sensors are shown in Figure 2.1. Such noise cannot easily be removed without severe degradation of the image, but can be minimised by keeping the sensitivity of the sensor as high as possible, so that the signal to noise ratio is consistently high throughout the chain of processes. The "speckle" seen in radar imagery is not strictly random, since it is an artifact of signal processing, and is related in intensity to the signal in the image. Systematic or regular noise is also found in satellite imagery, and can be largely removed during image pre-processing. The simplest form of systematic noise are line drop-outs, where a whole line, or part of a line, of an image is missing in the sense that the digital values are all zero (in some sensors, 255). This is due to a temporary breakdown on the satellite or in the downlink to a receiving station, resulting in a loss of signal. Most image processing systems replace these missing pixels by averaging the digital values in the lines above and below the missing data, giving an acceptable repair.

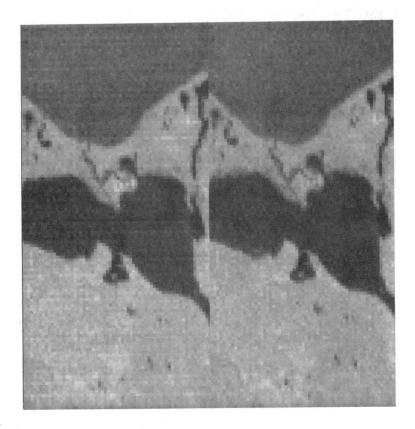

Figure 2.2. MSU-SK imagery of an area on the Baltic coast (courtesy Glavkosmos). The raw visible band image on the left has been de-striped using a statistical de-striping technique to produce the image on the right.

The most common type of systematic noise takes characteristic forms according to the sensor type. Image striping results from differences in sensitivity and gain between different elements of the sensor systems. Most optomechanical scanners, for example MSS and TM, use more than one array of sensors, so that the image is not generated one line at a time, but rather seven, eight or sixteen lines at a time. Each line of the sensor array will not usually have exactly the same sensitivity, and although gain is adjusted from time to time by the satellite operators, the correction is rarely perfect. This results in periodic horizontal striping in the image, repeated every seven lines in the case of Landsat MSS, and eight lines in TM images. The dynamic range of the striping is not normally very large, and the striping is often invisible in areas of the image which have a high dynamic range. It is most obvious over areas with low reflectance such as water, or in images acquired at relatively low levels of illumination. An example of striping in Soviet MSU-SK imagery, before and after processing, is shown in Figure 2.2. Horizontal striping is normally removed using a statistical de-striping technique, which sets the average for each line of pixels within the cycle of seven or eight lines, depending on the sensor, to the same value, correcting each pixel accordingly. Since it would be computationally expensive to take the average across the whole width of the image, and would also produce undesirable side-effects in high-contrast images, it is usual to compute the average progressively over a window tens or perhaps hundreds of pixels wide. It is most important to carry out this de-striping process before any geometric transformation of the image, since geometric correction will usually partially rotate the original image, and horizontal stripes will become diagonals, which are computationally much more difficult to remove.

Pushbroom scanners, such as those on SPOT, MOS-1 and MSU-E, use a very large single array of sensors, and individual charge-coupled devices within these arrays will rarely have exactly the same sensitivity. This results in vertical striping in the raw image, without the periodicity found in optromechanical scanner imagery. The earth rotation correction applied to most raw imagery at the receiving station transforms this vertical striping to a diagonal striping offset about seven degrees from the vertical, and renders it more difficult to remove. Most image processing systems now have the capability of de-striping pushbroom scanner imagery, even after earth rotation, although the results are often less satisfactory than for horizontal striping. Figure 2.3 shows part of a MOS-1 MESSR visible wavelength image, before and after de-striping. As with the optromechanical scanners, striping is most obvious over homogeneous areas of low reflectivity, and in winter imagery. The higher signal to noise ratios at infrared wavelengths compared with visible results in pushbroom striping being much less obvious in infrared images than in visible ones, as is illustrated in Figure 2.4.

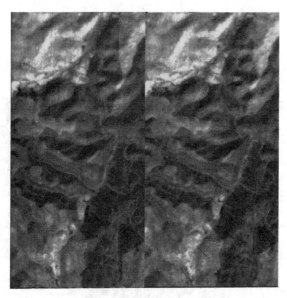

Figure 2.3. Part of a MOS-1 MESSR visible band image of an area in the Southern Uplands of Scotland (courtesy NASDA). The left hand extract shows vertical striping characteristic of pushbroom scanners. This has been almost removed using a de-striping technique, as shown on the right.

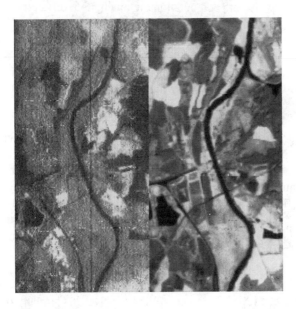

Figure 2.4. Visible (left) and near infrared (right) extracts from MSU-E imagery of the river Elbe in Germany (courtesy Glavkosmos), showing the higher dynamic range and reduced noise at near infrared wavelengths.

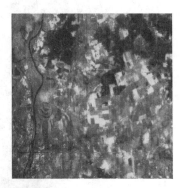

Figure 2.5. The use of Fourier filtering to minimise noise in digital imagery. A very noisy visible band MSU-E image of the Elbe river in Germany (top, courtesy Glavkosmos), when transformed to the Fourier domain (middle) shows systematic vertical and horizontal noise due to the pushbroom scanner and line drop-outs. A Fourier filter results in the less noisy image shown at the bottom.

Apart from the statistical de-striping techniques described above, it is possible to remove systematic noise using Fourier filtering techniques. The image is transformed to the frequency domain using a fast Fourier transform, and a magnitude image generated. Systematic noise can be clearly seen in such an image, and a filter is designed which then modifies the magnitude image before an inverse transform. Although computationally intensive, and not available on all image processing systems, Fourier filtering techniques can achieve superior results to statistical de-striping. This is especially true where the systematic noise is geometrically complex (for example, the noise is oblique to scan-lines), and Fourier filters can be designed to remove a wide range of defects in a single operation. A very noisy image from the Soviet MSU-E scanner is shown in Figure 2.5 (top). This image has along-track pushbroom striping, some line drop-outs, and a high level of random noise. A magnitude image resulting from a fast Fourier transform is shown in Figure 2.5 (middle), the three types of noise being highlighted. A filter can be designed to suppress the two varieties of systematic noise (pushbroom striping and line drop-outs) and also to emphasise the highest frequencies (edges, lines etc.) at the expense of the medium frequencies (mainly random noise). The effect of applying this filter to the real component Fourier image, and then re-transforming real and imaginary components into a cleaned image is shown in Figure 2.5 (lower). The pushbroom striping has vanished, line drop-outs have been reduced although not eliminated, and random noise is less obvious.

2.1.4 Radiometric Correction

Once sensor and transmission defects have been removed from the imagery or minimised as much as possible, further pre-processing stages may often be applied before the vital processes of image enhancement and information extraction start. Radiometric and atmospheric correction may be carried out, and the imagery may be geometrically corrected. Two images of the same area acquired on different dates may have very different digital values. This can be due to differences in solar elevation, depending on the time of year, differences in the amounts of atmospheric scattering and absorption, and finally to changes in land cover in the area studied. The last factor is almost always the one which remote sensing seeks to discover in studies of multi-date imagery, although in exceptional circumstances investigators might wish to use remote sensing to monitor atmospheric pollution, and might correct for solar elevation and land-cover to arrive at a measure of atmospheric scattering and absorption. For the purpose of this book we will assume that the land cover information is required, and that the effects of solar illumination and variations in the atmosphere should be removed as much as possible.

Radiometric correction to compensate for sun elevation differences between image dates and for differences in sensor calibration is an essential precursor to the detection of change by a comparison of reflectance values, but this is fortunately a relatively easy process. Since most earth-observation satellites pass over each point on the earth at roughly the same local sun time, solar illumination variations are mainly a function of the season. The solar elevation and azimuth at the time of image acquisition are normally recorded in the header of the digital image, and it is a relatively simple matter to adjust the digital values of a series of images to a constant solar elevation so that all images appear to have been acquired at the same time of year. Most scanners used in the minerals industry have a relatively narrow angle of view, and solar illumination can be assumed to be constant across the whole of the image. A simple correction for solar illumination is based on the fact that illumination varies with the cosine of the solar elevation angle, and imagery can thus be corrected for an arbitrarily fixed elevation.

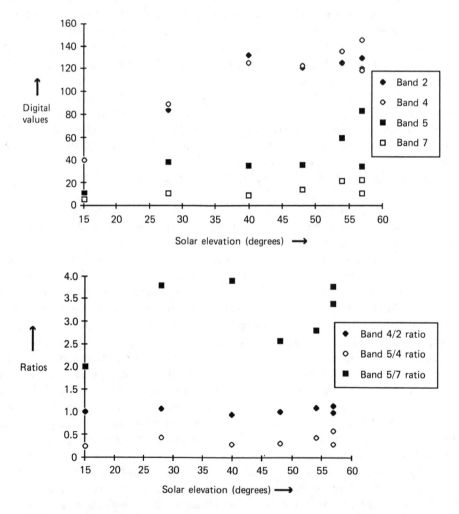

Figure 2.6. An example of the use of band ratios to minimise the effects of differences in solar elevation. Raw digital values (top graph) increase with increasing solar elevation, while band ratios (lower graph) are relatively unaffected. Landsat TM imagery of Portworthy mica dam, Lee Moor, south Devon.

Band ratios can eliminate most of the effect of solar elevation, as shown in Figure 2.6. More precise corrections also take into account the variation in earth-sun distance through the year, although for most purposes this effect is so small as to be insignificant. With some airborne scanners, for example the Daedelus, and for imagery from meteorological satellites, radiometric correction is more complicated. The wide scan angle of these systems leads to significant solar elevation differences, or differences in the angles between the sun and sensor, across the width of the image, and these may require compensation prior to any quantitative analysis. In its simplest form, the correction is an arithmetic one, based on a simple model of the scanner geometry, although more precise correction may require incorporation of bidirectional reflectance and even terrain information. These radiometric corrections can remove differences due to solar elevation and to the relative positions of pixels within broad scan lines, but they do not, of course, compensate for illumination differences due to topography. The use of precise digital elevation models, preferably with a resolution at least as fine as the satellite imagery, allows an illumination correction image to be generated, in which each pixel has a value corresponding to the relative solar illumination at that point. This image can then be used to correct the digital values in a co-referenced satellite image in order to remove the effects of topography. It is fortunately rare that such time-consuming correction is required in geological remote sensing.

2.1.5 Atmospheric Correction

Atmospheric correction is a greater problem, and there are a range of techniques of varying complexity to achieve this. It is fortunately the case that atmospheric effects are usually relatively minor in comparison with solar elevation effects, and most remote sensing for the minerals industry can be carried out without the need for atmospheric correction. The removal of atmospheric effects from visible wavelength data over water is however very important. Atmospheric effects contribute up to 80% of the signal reaching the satellite over water, and are a complex mixture of scattering and absorption by gases, aerosols and solids. A common approach is the "darkest pixel" method, where the water leaving radiance in the near infrared is assumed to be zero and the signal received at the satellite is thus entirely due to atmospheric effects. The effects of air molecule scattering can be calculated and removed, leaving the radiance due to aerosol scattering. It is then possible to produce ratios relating the effects of aerosol scattering in the near infrared to the aerosol scattering at other wavelengths, and thus to remove atmospheric effects from visible wavelength data. The corrected radiance values can be related to physical properties such as concentrations of chlorophyll and sediments in the water. The "darkest pixel" technique can also be used in land studies, provided that an area of deep (and hopefully unpolluted) water is available somewhere within the study image.

Another empirical method for atmospheric correction is known as the regression intersection method (RIM). This depends on the fact that, for a uniform type of land cover, for example pasture or coniferous woodland, differences in digital values between pixels in the same area of a single image will result mainly from topographical variations. If the digital values at two wavelengths for two different types of land cover are plotted on a graph, the intersection of the two regression lines will give a measure of the atmospheric contribution at each wavelength. A correction factor can then be derived for each of the six reflected wavelengths of the Landsat TM, for example, and all raw digital values corrected accordingly. Two severe problems with this technique are, firstly that few land cover types are actually very homogeneous, and secondly that it is rare to find a sufficient range of slopes and aspects for two distinct land-cover varieties within a single study area to give reliable regression lines.

A range of sophisticated atmospheric correction models, some of them incorporating meteorological data, have been developed for more precise atmospheric correction than is possible by simple empirical techniques. The LOWTRAN and TANRE models are widely used examples. Few applications in the mineral industry require such precise correction, however, and this is not the place to go into further detail on the subject.

2.1.6 Geometric Correction

The final stage of pre-processing of remotely sensed imagery, and one that must be preceded by most other pre-processing, is geometric correction. The reason for this is that geometric correction involves an irreversible modification of the pixel values of the image during re-sampling, as the original image is warped to fit a new coordinate system. Any correction for scanner defects or scanner geometry, or the relative geometries of scanner and incoming solar radiation must be done before this warping, although corrections affecting the whole scene identically, such as solar elevation correction or atmospheric correction in images covering a relatively small area of the earth's surface, can be carried out either before or after re-sampling.

There are two main routes by which geometric correction may be achieved, and in practice the two are often used together. The first technique depends on a precise knowledge of the orbit of the satellite, and of its position at the time of acquisition of the image. If the precise location, altitude and direction of pointing of the satellite are known, then three-dimensional trigonometry can be used to calculate the geographic latitude and longitude of each pixel in the image. This information is then used to transform the image to a chosen map projection. Modern satellites are extremely stable platforms, and information on their position and attitude is normally transmitted back to earth with the image data, enabling fairly precise geometric correction using suitable software. The second technique, which was refined during the early years of remote sensing when satellite ephemeris data was sparser and less accurate, is based on the identification of points of known latitude and longitude (ground control points, GCP's) within the image. A transformation matrix is then established to convert image coordinates to geographic coordinates. Given sufficient GCP's, this correction can be more accurate than the ephemeris technique, but is more computer intensive and usually requires greater human intervention, even when the recognition of GCP's in the imagery can be automated. The best transformation is usually achieved by a combination of the two techniques, with an initial correction using ephemeris data, and the use of a small number of GCP's to check and refine the ephemeris information. Both techniques assume flat topography, and do not achieve perfect image transformation in areas of extreme topography. This is especially true of sensors with a wide field of view, or sensors such as SPOT or synthetic aperture radars which can view an area from an oblique angle. Precise geometric correction in such cases requires the use either of digital elevation models or of stereoscopic imagery to correct for altitude variations.

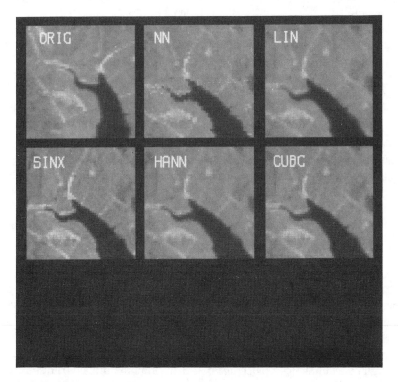

Figure 2.7. The effects of different re-sampling algorithms on image fidelity. The original image (top left) has been re-sampled using nearest neighbour, linear interpolation, sinx/x, Hanning and cubic convolution algorithms.

Once a transformation matrix has been established by one of the techniques described above, the actual re-sampling of the image must take place. In essence, this means that each digital value in the original image must be moved to a new position in the new corrected image. If all the moves are by a whole number of pixels, the process is relatively simple, and the digital values in the re-sampled image are unchanged, but merely re-arranged. Unfortunately, in a real image few pixels represent uniform surfaces, but most have a digital value that results from a mixture of different reflecting surfaces within the area of the pixel. It is also the case that most pixel moves within a transformation will not be by precisely integer amounts. If an image is re-sampled by this simple process, which is known as nearest neighbour re-sampling, the raw pixel values will be retained but the re-sampled image will have a very jagged and blocky appearance. A range of different techniques have been developed to achieve more precise spatial re-sampling while retaining as closely as possible the original digital values. The details of these techniques need not concern us here, but some brief comments may assist those who need to decide on a re-sampling technique for a specific project. Bilinear interpolation results in a less blocky image than nearest neighbour techniques, but often has unacceptably large modification of digital values. A whole family of algorithms based on what is known as sinx/x interpolation are offered by different organisations. Some of them produce excellent results, but are computationally very expensive, while those that are simplified to reduce computing time sometimes introduce undesirable artifacts to the image, in particular a kind of edge enhancement along boundaries between spectrally dissimilar surfaces. Another family of techniques are generally known as cubic convolution algorithms. These are similar in some respects to sinx/x, but their simplified forms usually result in less spectral degradation of the image. There is usually a trade-off in re-sampling performance, in terms of spectral and spatial fidelity to the original image, and cost. The cheapest technique, nearest neighbour re-sampling, produces a spectrally faithful but spatially poor image, while some more complex algorithms produce an aesthetically very pleasing image at the cost of severe spectral distortion. The customer should be aware of these problems, and should always request details of interpolation algorithms used by a contractor, as well as asking for alternatives. The effects of using different re-sampling algorithms are illustrated in Figure 2.7.

2.2 SECOND STAGE IMAGE PROCESSING
The second stage of image processing can itself be sub-divided into two main groups of activities, image enhancement and information extraction. The first of these includes a diverse range of techniques to improve the appearance and interpretability of digital imagery, and is directed particularly at a final product which is going to be subject to visual interpretation. The second uses the digital nature of the imagery, and the processing power of the computer, to extract information, either in the form of new images or maps, or as tables of statistics, required by the user, often minimising the requirement for skilled interpretation of the image by a human observer.

2.3 IMAGE ENHANCEMENT
2.3.1 Contrast stretching
Most image processing systems have a dynamic range of eight bits per channel, with grey tones ranging from black=0 to white=255. Satellite images, particularly those at visible wavelengths, often have much lower dynamic ranges, especially at times of low solar illumination. For better interpretability, either on the monitor of an image processing system or in photographic hard copy, the original digital data need to be transformed to occupy the full dynamic range of the system. This process is known as contrast stretching. An example of an image before and after contrast stretching, together with histograms showing how the distribution of digital values changes, appears in Figure 2.8. Contrast stretching can be done manually, usually allowing the operator to interactively move the upper and lower limits of the digital numbers, and to

move the position of the mean. In some systems, the stretch can be carried out in multiple parts. Most image processing systems also have a variety of automatic contrast stretches, based on the generation of gaussian, equalised, or other types of standard distribution. Contrast stretches are normally applied initially through a look-up table; the original data are not themselves modified in the frame store, but a computation based on a table of values is made between the store and the display device. Once a satisfactory stretch has been achieved, then the original numbers can be transformed for production of hard copy.

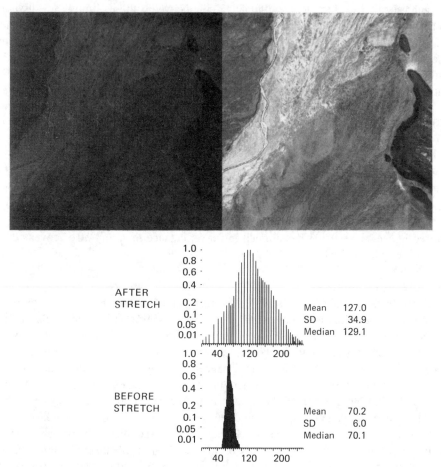

Figure 2.8. The effect of contrast enhancement. The original image (a Landsat MSS extract of part of Northern Kenya, courtesy EOSAT) has a low dynamic range, as illustrated at top left and in the lower histogram. Contrast enhancement expands the digital values to fill the dynamic range of the display system, producing the image shown at top right, and the upper histogram.

2.3.2 Density Slicing

The human eye can perceive subtle differences of colour more clearly than slight differences in monochrome grey level. In order to enhance differences in single-band images, it is often useful to assign a colour to each grey level, or to groups of digital values. This process is known as density slicing. A simple density slice assigns a uniform colour to each user-defined range of grey levels, enhancing the between-group differences but completely suppressing within-group variation. More sophisticated density-slicing procedures allow the user to set colours for the upper and lower limits of each slice, and individual digital values between these limits are assigned intermediate colours. This still achieves the purpose of highlighting gross differences and similarities, but preserves detail in the image. The difference between the two processes is illustrated in Plate 2, a, b and c. Plate 2a is actually a digital elevation model, rather than a satellite image, since this has a large dynamic range with smooth spatial variations in digital value, and illustrates more clearly the differences between the processes. This grey level image is fairly difficult to interpret, and it is particularly difficult to judge whether two hills (light colour, high digital value) are of the same elevation. A simple density slice (Plate 2b) produces the effect of a contour map, with solid colours for each contour interval, but does not preserve any detail within each interval. Plate 2c, using a more complex slice, preserves detail at the same time as colour-coding according to height. Apart from aiding the interpretation of single-band images, density slicing can be used to prepare masks for separation of selected areas of a satellite image. For example, water areas can be separated by density slicing an infrared image, since water has a much lower reflectance than any other cover type at this wavelength.

2.3.3 Colour composites

Since most satellite imagery is available in multi-band formats, the examination of the data one band at a time does not extract the maximum information. Inter-relationships between different wavelengths are very important in the recognition of features and cover types, and it is useful to display more than one band simultaneously on the image processing system, and to produce multi-band hard copy. This is often done through the use of colour composite images, where three bands of data are assigned to the blue, green and red colour guns of the colour monitor. The number of ways in which multi-band imagery can be displayed in three colours rapidly becomes very large. For a three-band data set such as SPOT-XS, there are six different possible colour combinations, but for the seven bands of Landsat TM there are eight hundred and forty! In order that image interpreters can recognise features on imagery processed by others, certain conventional combinations have evolved. Three-band data sets with wavelengths approximating to visible green, visible red and near infrared (Landsat MSS, SPOT-XS, MSU-E, MOS-1. IRS-1) are conventionally displayed with band 1 (green) assigned to the blue gun, band 2 (red) assigned to green, and band 3 (NIR) displayed in red. This results in the familiar false-colour composite showing vegetation in red and water in blue. For land-use studies, Landsat TM has valuable information in the mid infrared (band 5), which needs to be included in a general-purpose colour composite. For this reason, the standard TM composite has band 3 (red) assigned to the blue gun, band 4 (NIR) assigned to red, and band 5 (MIR) in green. Typical examples of these two conventional colour composites are shown in Plates 1 and 3. Different colour composites are often required for geological mapping and mineral exploration, since the mid-infrared band 7 was specifically designed to be sensitive to clay minerals. Composites including bands 7 and 5 with one visible band (usually 3) are frequently used for this purpose.

In most satellite images the correlation between different bands of the sensor is very high. The generalised reflectance (albedo) of most surfaces, in particular rocks and soils, varies relatively little with wavelength between different surface types, and spectral differences are relatively subtle compared with the total reflectance. If correlation between bands can be minimised, spectral differences are emphasised, and the interpretability of the image is increased. The processes which achieve this are known as decorrelation stretching. In some cases, the original colour tones of the colour composite are preserved but emphasised, while in others a new set of more distinct colours replaces an originally small tonal range. The two main types of decorrelation stretching are based on principal components analysis and on intensity-hue-saturation transforms, both of which are discussed in more detail below.

2.3.4 Ratio Images

The inter-relationships between different bands of data may often be rather subtle, and techniques other than production of colour composites may be necessary in order to reveal them. It may also be desirable to remove some component, such as illumination due to topography or solar elevation, which is common to all bands of an image in order to emphasise more subtle features which differ between bands. Finally, multi-band data sets with more than three bands cannot be represented simultaneously in the colour composite, and some means of data reduction is often required. The arithmetic combination of different bands of satellite data to produce new composite bands often involves the use of ratios between bands, and so this group of arithmetic combinations are known collectively as ratio images. They can range from simple ratios (one band divided by another) through normalised ratios (the difference between two bands divided by the sum of the same two bands) to very complex ratios involving many bands. Their effect is always to subdue differences due to illumination, either topographical or inter-date, and the intention is to emphasise spectral differences between the chosen bands.

A commonly used ratio in land-cover studies is the Normalised Difference Vegetation Index (NDVI), given by the equation

$$NDVI = \frac{(Rnir-Rred)}{(Rnir+Rred)} * K$$

where Rnir and Rred are the digital values at near infrared and red wavelengths respectively, and K is a constant, usually a large number (128), in order to result in a ratio image with a large dynamic range. The NDVI is used as a semi-quantitative measure of vegetation density and vigour (more specifically, chlorophyll activity), and can be derived from a whole range of spaceborne sensors, from AVHRR (one kilometre resolution) to SPOT (twenty metre resolution). The comparison of a false-colour composite image with an NDVI image in Plate 4 illustrates the way in which this ratio highlights vegetation differences. Most ratios used in geology are less quantitative, although prior to the availability of Landsat TM data, ratios were developed for use with spectrally poor MSS imagery to quantify iron oxide contents in surface materials.

2.3.5 Principal Components

A mathematically more complex, and in some respects less controllable means of reducing a multi-band data set to a smaller number of more significant bands is known as principal components analysis. In essence, a series of n bands of original data are converted to n new bands, where the first principal component contains the information most highly correlated between all the original bands, the second principal component contains the next most highly correlated data and so on to the nth principal component, which contains the least correlated information. The first principal component always contains the main topographical features of the imagery, since this is the primary control on digital values. Which of the other principal components contains the most significant information on, for example, vegetation abundance or clay mineral alteration is very scene dependent, varying with the proportion of the scene actually occupied by that cover type. Despite this limitation, principal components analysis is a valuable means of data compression, and of producing a set of de-correlated images from an originally highly correlated data set. The use of directed principal components in the location of alteration zones is described more fully in a later section.

2.3.6 Convolution Filtering

The principle of convolution filtering needs to be explained, since filters are used in many different processes. They are particularly important in smoothing an image, in edge-enhancement, linear filtering, and in edge and line detection. Convolution filtering of digital imagery is actually fairly simple, but should be understood if the significance of these processes is to be appreciated.

Filters can be of various sizes, but the process can best be illustrated with a small 3*3 pixel filter, as shown below. This is actually a smoothing or averaging filter. The numbers in each square of the matrix indicate the weighting to be given to each pixel. The box is moved over the whole image, pixel by pixel, and a new image is generated. For each position of the box, a new value is generated for the centre pixel position of the box in the new image, based on the sum of the weighted values of all the pixels covered by the box in the old image, divided by the number of cells in the box, in this case nine. The new image will be smaller than the old by the size of the filter minus one pixel, since the filter cannot operate beyond the edge of the original image.

```
                    1  1  1
A Smoothing Filter  1  1  1
                    1  1  1
```

Original Image, represented as numbers (eg, brightness values)

```
100  100  100  255  100  100  100  50  50
100  100  100  255  100  100  100  50  50
100  100  100  255  100  100   50  50  50
100  100  100  255  100  100   50  50  50
100  100  100  255  100   50   50  50  50
100  100  100  255  100   50   50  50  50
100  100  100  255   50   50   50  50  50
100  100  100  255   50   50   50  50  50
```

Results of applying smoothing filter (filtered image is smaller than original by the filter size minus one)

```
100  152  152  152  94  78  61
100  152  152  152  89  72  56
100  152  152  146  78  61  50
100  152  152  141  72  56  50
100  152  146  129  61  50  50
100  152  14J  124  56  50  50
```

```
                      -1  0  1
Linear Gradient Filter -1  0  1
                      -1  0  1
```

Result of applying linear gradient filter to the original image shown above

```
100  152  100  48  95  83  89
100  152  100  48  89  83  95
100  152  100  43  83  89  100
100  152  100  37  83  95  100
100  152   95  32  89  100  100
100  152   89  32  95  100  100
```

These two simple examples illustrate the principle of convolution filtering. The technique is arithmetically simple but highly repetitive, ideally suited for computers. Most image processing systems allow kernel sizes of up to 7*7, and have libraries of pre-defined filter kernels for a range of tasks. A few systems allow larger kernels, but unless some kind of hardware arithmetic accelerator is provided, the penalties in computing time are usually not compensated for by the increased usefulness of the resulting imagery. Most essential tasks are accomplished using small filters, 3*3 or 5*5, which are relatively fast even on simple image processing systems. Very large kernels are important in specialised spatial analysis, and can also result in geometric transformation with a minimum of distortion of pixel values.

2.3.7 Edge Enhancement and Linear Filtering

If hard-copy imagery is to be produced for visual interpretation, it is often desirable to artificially "sharpen" the image, exaggerating linear features and boundaries between different cover types. The process known as edge enhancement achieves this by applying a high-pass filter to the image, and then adding the filtered image to the original. This results in an intensification of boundary and line information, although it can severely distort the individual pixel values near such edges. An example of a single-band image before and after edge enhancement is shown in Figure 2.9.

In geological structural studies, it is often desirable to enhance linear features on imagery using a low-pass filter which eliminates most of the surface detail at the same time as enhancing linear features. There are an almost infinite number of possible filters, and most image processing systems allow the users to define their own filters as well as to use a pre-defined library. The present author often uses a set of three very simple gradient filters with the following values:-

```
-1   0  +1      0  +1  +2     +1  +1  +1
-1   0  +1     -1   0  +1      0   0   0
-1   0  +1     -2  -1   0     -1  -1  -1
```

The resulting three images, emphasising linear features in mutually orthogonal directions, can then be combined either as a blue - green - red colour composite or as a single monochrome image. Artifacts produced by any single filter will not be reinforced by the others, and the resultant image is an approximation to a non-directional filtered product. An example of this type of filtering used to enhance geological structure in a coarse-resolution Meteosat image is shown on the cover of this book.

2.3.8 Colour Space Transformations

Colour images, whether displayed on the monitor of an image processor or in photographic hard copy, are normally composed of three bands of data, usually three different wavebands of an image but potentially three different co-registered images. Each band is assigned to one of the colour guns (red, green or blue) of the colour monitor, or colour lasers of the filmwriter. This is not, however, the only way in which a colour image can be made up. Any colour image can be described in terms of intensity, hue and saturation, instead of red, green and blue. Intensity is a measure of the brightness of each pixel, hue is a measure of the colour, and saturation measures the depth or purity of the colour. Many image processing systems allow the conversion of a red - green - blue image to intensity - hue - saturation colour space, as well as reverse transformations. These colour space transformations are used for two main purposes. Firstly, they can be used as an alternative means of de-correlation stretching. If an RGB image with a high degree of inter-band correlation is transformed to IHS, and the hue then stretched to fill the whole dynamic range before backward transformation to RGB once more, slight spectral differences between surfaces will be enhanced, although the colours of the de-correlated image will bear little or no resemblance to those of the original. An example of IHS de-correlation stretching of a Thematic Mapper scene is shown in Plate 5. The second main use for colour space transforms is in the combination of different data sets, as illustrated diagrammatically in Figure 2.10. Three totally disparate but co-located data sets can be combined into a single image as an aid to visual interpretation of their inter-relationships. This example, the result of which is shown in Plate 6, combines filtered satellite imagery (intensity), a geological map (hue) and a digital elevation model (saturation) as an aid to hydrogeology. As a general guide, the data set with the highest spatial information content (linear features, texture) should be assigned to intensity, and the data set with the highest dynamic range to hue. Small variations in saturation are not perceivable to the human eye, so this should be used for a less important data set with a broad dynamic range and preferably abrupt rather than gradational boundaries.

Figure 2.9. The effect of edge enhancement. Part of a SPOT PAN image of an area in Scotland (top, copyright CNES 1987) has been edge-enhanced to give the "sharper" but radiometrically distorted image shown below.

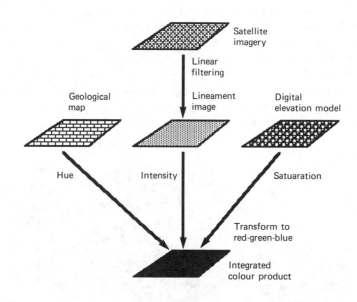

Figure 2.10. The use of IHS transforms for integration of disparate data sets.

2.4 INFORMATION EXTRACTION
2.4.1 Visual Interpretation
The human eye is a powerful tool for detecting subtle differences in texture and recognising characteristic shapes, patterns and feature associations. Automated classification techniques cannot always take into account the textural, contextual, pattern and shape information associated with image features which are essential for their recognition and differentiation. Simple texture measures can be derived from imagery and combined with spectral information for classification, but in many cases the accuracy of automated classification, of whatever type, is considerably poorer than that achievable by visual interpretation. This is especially true in the interpretation of satellite imagery for geological mapping.

Visual interpretation can be performed on transparent film overlaid on high quality photographic prints, directly onto lower quality paper plots or on digital imagery using interactive image processing equipment. A typical geological interpretation, together with the imagery from which it was derived, is shown in Plate 7. For many applications, working digitally or with high quality photographic products is preferable. Drawing digital boundaries using interactive image processing equipment for areas larger than the display screen size may be a time consuming operation and annotation of hard copy imagery will probably be preferable for large areas. If the interpretation is required in a digital form the hand-drawn overlay can easily be digitised using standard cartographic equipment. Interpretation is always best performed by an expert who is familiar with the geographical area covered and the features being interpreted and who has received basic instruction on the significance of variations in colours and tones in terms of the physical, structural and environmental characteristics of the features under study.

2.4.2 Classification

In principle, classification techniques use the spectral, and sometimes also the spatial, properties of digital imagery in order to sub-divide the imagery into meaningful classes of different cover types. Classification attempts to emulate the activity of the human interpreter, who sub-divides imagery enhanced by a series of appropriate processes into a series of classes, based on experience and on the requirements of the project. The human interpreter uses colour, texture and context in order to identify specific features and cover types. In theory, an automated classifier can examine many bands of data at the same time in a totally objective fashion, at a speed thousands of times faster than human interpretation. In practice, no computer based classification system is yet as accurate as an experienced visual interpreter, mainly because the three critical aspects of colour, texture and context cannot be evaluated equally well by most current image processing systems, and because the computer operates with a set of relatively inflexible rules, and cannot, in most cases, learn from its own mistakes.

Most common classifiers operate on the basis of "colour" alone, in the sense that they operate on the individual pixel values at each wavelength. Each pixel is assigned to a class, feature or cover type based on its own spectral properties, without any consideration of surrounding pixels. These "per-pixel" classifiers can be divided into two main groups, the unsupervised and supervised classifiers. Unsupervised classification demands no prior knowledge of the image, but effects a sub-division based on the intrinsic properties of the digital data. In principle, the digital values for each pixel at each wavelength are examined, and the image then sub-divided into a pre-set number of classes. Some image processing systems allow the user to select the number of classes, or the amount of spectral difference that there should be between classes, while others are less flexible. In practice, the initial clustering process to identify the main natural classes in the image usually examines only a sample of the total number of pixels in order to save time. Once class statistics have been established, the user can normally select the number of classes required in the final image, based on tables and sometimes graphical representations of the spectral distribution of the clusters. The final classification of the whole image is then undertaken, normally using maximum likelihood techniques (see below) to assign each pixel to its most appropriate class. Provided that the process is reasonably rapid, it is often useful to carry out unsupervised classification of an image, in order to obtain an objective impression of main spectral types, before attempting any supervised classification. Some complex natural cover types, such as semi-natural upland vegetation, may sub-divide better on the basis of unsupervised clustering due to the difficulty of defining reliable training areas for supervised classification.

Supervised classification depends on some prior knowledge of the area covered by the imagery to be classified. The operator interactively informs the system that a particular group of pixels in the image represents a specific cover type on the ground, and the computer then searches for pixels of similar spectral characteristics. The areas defined interactively are known as "training areas", and much has been written about techniques for selecting appropriate training areas, and about the amount and quality of "ground truth" required for satisfactory classification. The main essentials are that the training area should be as homogeneous as possible, and that it should be representative of the class which the system is required to identify. Once a group of training areas defining the main cover types of interest in the specific study have been selected, then classification can proceed. Most image processing systems allow at least two types of supervised classification. A simple box or parallelepiped classifier defines the mean and range of digital values for each class at each waveband, based on the training areas, and then classifies all pixels lying within these ranges as belonging to the class. Often only one class can be classified at a time, and there is inevitably considerable overlap between classes, unless the training areas are so homogeneous as to be unrepresentative of real-world surfaces. A more sophisticated method of classification uses statistical techniques to assign each pixel to a class. Minimum distance and maximum likelihood classifiers use such approaches, and can usually handle a large number of classes simultaneously. Some image processing systems allow interactive adjustment of probability levels in order to achieve an acceptable classification.

Classification techniques are widely used for land-cover mapping studies, but do not find wide application in geological mapping. Classification accuracies in lithological studies are usually very low, except where vegetation is scarce or absent, and topography subdued. Although a very precise geological interpretation can often be made of suitably enhanced imagery by an experienced interpreter, automated classification of the same imagery normally fails. The human observer depends heavily on textural and contextual information in such interpretation, and also applies geological concepts such as directional continuity of marker horizons, the probability of symmetry about fold axes and the effects of faulting, which assist and guide the interpretation. In the context of the minerals industry, automated spectral classification is of value mainly in land-cover mapping for planning or environmental impact assessment purposes, and not for automated production of geological maps.

On the basis of five years experience of a wide range of remote sensing applications projects at the UK National Remote Sensing Centre, it was concluded that classification accuracy from per-pixel multispectral classifiers ranges from 60 - 80 %, depending upon cover type and sensor. Accuracies appear to be higher from Landsat TM than from SPOT, despite the slightly lower spatial resolution, and resulting from the additional spectral channels available with TM. Accuracy is higher for spectrally homogeneous cover types displaying low intra-class variability. Classification can be improved by using multi-date imagery, but this is normally limited, in humid temperate climates, by data availability due to cloud cover. Unsupervised clustering is more appropriate in areas of complex semi-natural vegetation, but good quality ground data is needed to interpret the results. Accuracy can be significantly improved by pre-classification segmentation of imagery in order to exclude unwanted areas from the classification. The image can be segmented by altitude, using digital elevation models or a digitised contour mask, or by use of intrinsic properties of the satellite data.

2.4.3 Change Detection

The detection of change over time is one activity that can be performed more efficiently with digital remotely sensed imagery than by almost any other means. The relative ease with which different dates of satellite imagery, separated in time by days, months or years, can be co-registered with each other, and the combination of quantitative measurements of surface reflectance at a number of wavelengths permits automated detection of change. The kinds of change of interest are extremely varied. International agencies and governments may wish to monitor removal of forest or the encroachments of deserts. Mining companies may wish to monitor changes in their own or their competitors' surface pits and dumps. Mine planners may wish to observe which areas around a proposed new mine site are subject to the highest rate of change in terms of conversion of forest to agriculture or even desert to irrigated land, so as to site mine workings and townships in the least sensitive areas. Change detection can even be used to monitor soil erosion around mines or along their transport routes.

A change in digital values between two dates of imagery of the same area can result from real change of surface cover, or from atmospheric or solar elevation differences. If a precise comparison of radiance values at different wavelengths is required, then the imagery must be corrected for solar elevation, as described in a preceding section, and some kind of atmospheric correction must also be applied. It is usually simpler to use ratios, which are almost independent of solar elevation, and less affected by atmospheric differences than raw single-wavelength values, in order to highlight areas of change. The NDVI, or a similar ratio of infrared to red reflectance, is commonly used for this purpose, since changes in vegetation (most usually from vegetated to non-vegetated, but sometimes the converse) are either the actual change being sought, or are indicative of it. A particular case of change, and one which is subject to little ambiguity, is change in areas of surface water. The very low reflectance of water at infrared wavelengths compared with visible makes it the most easy surface cover class in which to detect change. Once change has been detected, either in corrected imagery or in ratio images, the significant change has to be separated from change which is of no interest in the context of the study. In most areas of the world apart from tropical rainforests and deserts, seasonal variations in vegetation amount and distribution are the most obvious change visible in satellite imagery. In many parts of the world the differences between "dry season" and "rainy season" vegetation distribution are extreme, while in more temperate climates the complex changes in cropping patterns due to crop rotation and variations in planting and harvesting dates for different crops can result in complex patterns of change which can obscure the more specific change of real interest to the investigator. This spurious change can be minimised if imagery from the same month, but different years, is used to assess change between years. The problems with the use of "anniversary" imagery are that one year is rarely precisely the same as another in terms of the precise timings of vegetation activity because of inter-annual variations in rainfall and temperature, and also that, especially in temperate maritime climates, it is very difficult to be sure of acquiring cloud-free imagery at the same time each year.

In some cases the best approach, if inter-annual change in, for example, the extent of urban areas is required, is to use imagery from roughly similar seasons. Comparison of NDVI images, subtracting one date from the other for example, highlights areas where change has occurred. For studies of relatively small areas, the areas of change can be inspected visually on the image in order to assess the type of change. For larger areas, where visual interpretation might not be appropriate, a series of different ratios, selected on the basis of the known or inferred spectral characteristics of the surfaces of interest, can be used to further sub-divide those areas in which change has occurred.

An alternative technique is to classify each date of imagery into the land-cover classes of interest, using whichever classifier is found to be most appropriate. The classifications from each date are then compared digitally, and a change matrix generated showing percentage change from each cover class at one date to each class at the second date. The problem with this technique is that classification errors, which vary with cover type but are always present, are additive between dates of imagery. The change statistics will thus include the errors of all cover classes at all dates. For classes which can be reliably mapped (water and woodland, for example) this technique has definite advantages in that it can give a direct quantitative measure of change, but it should not be used for classes where the classification accuracy is less than about 85%.

A change image is shown in Plate 8, an example of change detection using classified images. The areas of mineral workings at each date have been classified in the original TM images of the mine sites, and the two classifications then compared digitally. This is obviously a very simple example, with only one land-cover class, but the process could, in principle, be applied to complex multi-class images.

2.4.4 Pattern and Proximity Analysis

An exciting possibility of image processing, and one that has so far been little exploited by manufacturers of image processing systems, is the ability to derive quantitative measurements of patterns of distribution of surface features and their proximity to each other. These measurements can be very useful in assessments of environmental impact, in choosing between alternative routes for roads, railways, pipelines and similar features, and in numerous other aspects of planning. They can also be used to model geological features favourable for the occurrence of mineral deposits, as will be shown in a later section. Spatial analysis allows GIS operations to be carried out on raster data sets derived from satellite imagery and other co-registered data.

In its simplest form, the analysis of pattern deals with a set of features or areas with a single attribute, for example surface water bodies, lakes, ponds and rivers. Most image processing systems allow the measurement of the total area occupied by this cover type. Some will provide the total perimeter length of the cover type. A few systems allow the elements of a single class to be sub-divided into a user-defined number of groups according to their area, producing statistical tables of size distribution as well as a new image where the members of each sub-class are numerically coded. Figure 2.11 a and b shows how this type of sub-division can be carried out. Sub-division of elements of a class according to their shape is possible on very few remote sensing image processing systems, although this feature has been available on some metallurgical image analysis systems for many years.

The analysis of pattern, in terms of size and shape distribution, of more than a single class, is not possible on commercial remote sensing image processors, although this is also possible on some image analysers designed for microscopic investigation. There is obviously a case for better inter-disciplinary communication.

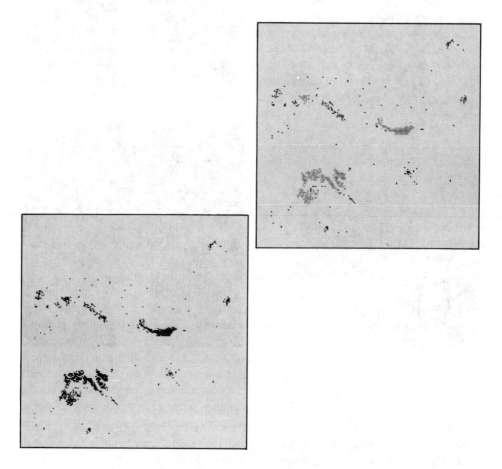

Figure 2.11. An example of pattern analysis. The illustration on the left shows areas of woodland derived by classification of Landsat TM imagery. Size analysis of the classified image produces the size category image shown on the right.

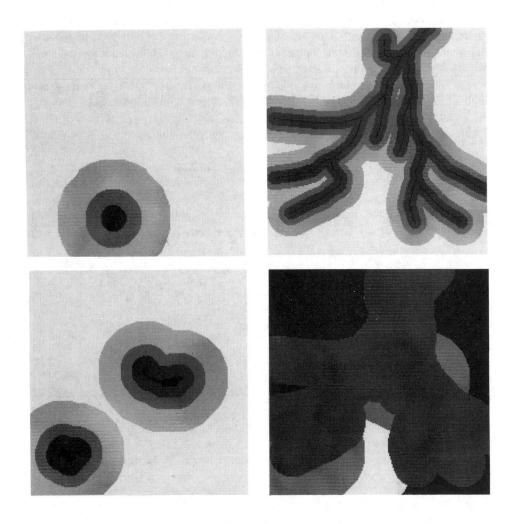

Figure 2.12. An illustration of proximity analysis. A tailings dam must be sited close to the concentrator, but far from rivers and protected ancient woodlands. An image of proximity to the concentrator (top left) is combined with proximity to rivers (top right) and ancient woodlands (bottom left) to indicate the permitted area for the new dam (white in lower right illustration).

Proximity analysis is carried out by generation of corridors or buffers around elements of a selected cover class. This is possible on many systems. A new image is generated, each pixel having a value dependent on its distance from the nearest occurrence of a specified cover class. These values indicate the proximity of each pixel to the selected cover type. If the proximity of one cover type to another is required, a mask of the second cover type can be superimposed on an image of proximity to the first. Multiple proximity images can be generated for environmental sensitivity studies. If, for example, it has been specified that a new tailings dam should not be closer than 4 km to ancient woodlands, or within 2 km of any surface water courses, but should be within 5 km of the mine concentrator, the superimposition of three proximity images, one for ancient woodlands, another for surface water, and the third for the concentrator, would permit the use of a simple box classifier to identify those areas which meet these specifications. The location of surface water and woodland areas could be derived from remotely sensed imagery, while the position of the concentrator would be accurately known, and could be digitised as a single point. Figure 2.12 a, b, c and d illustrates this process. This is a very simple example, but illustrates the possibilities for this kind of relatively simple spatial analysis.

2.4.5 Line and Edge Extraction

Although most structural analysis of remotely sensed imagery is carried out by visual interpretation, sometimes of specially enhanced imagery, this is a very subjective process, and depends for its success on skilled interpreters. There is also the problem that the results of such visual interpretation are then in analogue form, which may require digitising in order to carry out quantitative analysis or prior to integration with other data sets such as geophysical maps. It would be an advantage in some cases to have a quantitative automated means of extracting linear features from satellite imagery, although the limitation of automated line and edge detectors is that all linear features, whether of geological significance or not, are extracted. The human interpreter will usually tend, consciously or unconsciously, to ignore obviously man-made lineaments such as field boundaries and railway lines. The most effective edge and line detectors (the distinction between a line and an edge, in image processing terms, is that a line is a fairly narrow (two or three pixels) linear feature of distinctly different reflectance from the material on either side of the line, while an edge is the boundary between two relatively large homogeneous areas of different reflectivity) produce two derivative images, magnitude and direction, from each original image. Each pixel in the magnitude image has a digital value indicating the strength of line or edge, in terms of reflectivity contrast, in the original image. Pixels in the direction image are numerically coded according to the direction of the line or edge detected at the pixel, with eight or sixteen direction classes. Both magnitude and direction images will usually have values of zero if no edge or line is detected at that point. Line and edge detectors are essentially convolution filters. There are many filters which have been designed for this task, and details can be found in most standard image processing texts. As an example of the types of product which can be obtained from edge detectors, Figures 2.13, a, b, c and d show, respectively, an original single-band

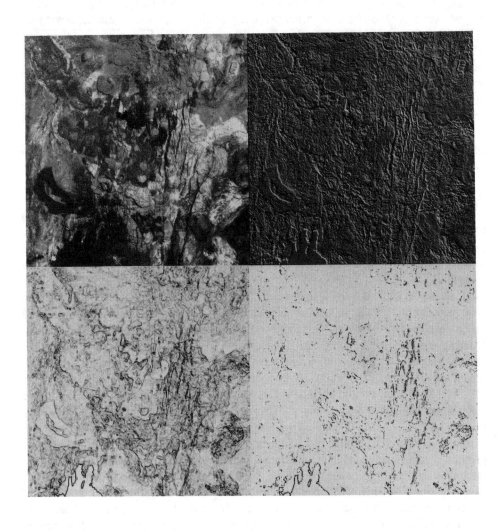

Figure 2.13. An illustration of automated edge detection. See Section 2.4.5 for explanation.

satellite image, a filtered version designed for visual interpretation, a magnitude image resulting from a edge detector, and an image of only the most intense edges. These images can then be further processed to provide more useful information, from an exploration point of view, about the structure of the area. If a digitised geological map was overlain on the filtered image, the structural elements could be studied for each major stratigraphical or lithological unit. Further processing of the magnitude image can result in derivative products showing lineament density (important as an indicator of fracture density in mineral and hydrocarbon exploration) and lineament length by analysis of connectivity between pixels and groups of pixels. Lineament density can be derived by smoothing the magnitude image, in order to average line magnitude over pre-determined windows, followed by density slicing. If used with caution, edge and line detectors can provide extremely useful quantitative information for geological mapping and mineral exploration. It must be emphasised, however, that this is no more a "stand-alone" technique than is visual lineament interpretation, and that analysis of the results must be carried out by structural geologists with knowledge of both the geology and physiographical features of the area under study.

2.5 IMAGE PROCESSING SYSTEMS

Operational use of remote sensing is becoming more widespread for a whole variety of reasons, one of the most important of which is the growing accessibility of high-performance image processing systems. In 1980 an image processing system capable of carrying out most of the operations described in this book would have cost about quarter of a million pounds sterling, and would have included a fairly large minicomputer requiring a large specially equipped and climate-controlled computer room and specialist staff. In 1990, image processing operations could be carried out on a system based on a "super-PC" type of personal computer, operating in a standard office environment without the need for highly specialised staff, and costing of the order of twenty-five thousand pounds. Although the size and cost of image processing systems has decreased remarkably, their basic structure remains fairly constant and is illustrated in Figure 2.14. The main components are as follows.

2.5.1 Data Input

For remotely sensed imagery, the overwhelming majority of image processing systems currently (1991) use tape drives. Digital imagery has historically been supplied on computer compatible tapes (CCTs) by the receiving stations, and this has proved a reasonably transportable and durable medium. Optical disks are ideally suited to remotely sensed image data, which needs to be written once and read many times, without any modification to the original data, but there are as yet no internationally accepted data format standards, and this is not yet a common medium for data supply. Most systems now use a 6250 bpi (bits per inch) tape drive, which is faster in use and requires a smaller number of tapes than the older 1600 bpi systems. Floppy disks are used for data input on some small image processing systems, although the relatively small data volume compared to the size of typical satellite images restricts their use in serious operational work. Tape cassettes, Bernuolli drives and other high volume rapid retrieval devices have been tried but are not as yet standard. Analogue spatial data, such as maps and photographic imagery, can be input via video camera systems or through digital scanners. Other map data can be input with greater precision using digitising tablets. Within a single organisation, data can be input into an image processing workstation through a local high-speed network, although the volume of data involved in satellite imagery does not usually permit the use of commercial regional or international networks at their current speeds.

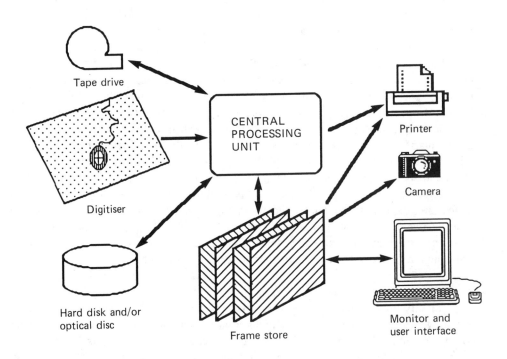

Figure 2.14. Schematic layout of a typical image processing system.

2.5.2 Data Storage

All image processing systems use hard disks to store the imagery which is currently being processed on the system, although there is a growing trend towards the use of small optical disks for on-line storage, despite the lower data transfer rate than is generally possible with hard disks. Some systems now achieve very high transfer rates to and from hard disks, minimising one of the serious bottlenecks restricting system speed. There is a growing use of large solid-state memories to hold images undergoing processing, as well as to provide very large virtual image stores, and although such memory is currently expensive, the gain in processing speed and flexibility is impressive, and prices are reducing.

2.5.3 Central Processing Unit

In the simplest configuration of image processing system, this is the host computer on which all the numerical operations required in image processing are carried out. It can range in size and power from a supercomputer down to a very simple personal computer, or even a single chip. In practice the configuration is rarely this simple. Many image processing workstations have two central processors, one actually in the workstation and the other in the host computer. Duties are allocated between the two according to the relative power of the two processors. If the workstation has a powerful processor, then the host will be used only for input/output activities, while a simple, relatively unintelligent workstation may make much greater use of the host. Many more powerful systems have special processors to carry out particularly demanding parts of the processing. Array or pipeline processors can carry out multiple numerical tasks such as convolution filtering or re-sampling, easing the load on the central processor and speeding up the whole operation. There is increasing use of transputers as accelerators in image processing, and groups of transputers can sometimes be added in a modular fashion to workstations to increase their operating speed.

2.5.4 Framestore

Sufficient random access memory must be provided to store the entire image to be displayed by the image processor. Until recently, most systems were restricted to an image display of 512 by 512 pixels, although larger displays are now becoming more common. In order to display the full dynamic range of three bands of a multispectral satellite image, and to allow for additional material such as classifications and map information on overlay planes, the frame store should be at least 28 bits, and preferably 32 bits deep, allowing for three 8-bit images plus overlays. For even the simplest systems, this implies one megabyte of rapidly accessible RAM. In almost all image processors, this is provided as a separate graphics board (for PCs) or frame store (for minis and upwards) which not only stores the displayed image but also includes some image processing functionality and the means of generating video output and addressing display devices.

2.5.5 Display Device

In almost all systems, this is a high-resolution colour monitor. The size will depend on the working environment and the preference of the user, although the physical parameters of the display must match the frame store. Relatively long persistence monitors are normally used to minimise flicker, and thus operator fatigue. A flat screen is also preferred to minimise image distortion. The brightness and contrast settings on the monitor should be calibrated to give an acceptable colour balance on a test image of some kind, and then should be sealed so as to prevent inadvertent re-setting.

2.5.6 Hard Copy Output

This can be the most expensive part of an image processing system, if high-quality hard-copy is required. Photographic output from a whole satellite image, rather than just the portion of the image displayed at any single time on the image processing monitor, must be generated using a film writer. These are high-precision devices that write very fine-grained film negatives or positives from pre-processed data tapes. Most of the modern generation of film writers produce a colour image at a single pass using colour lasers, although some older systems which produced a separate negative for each colour band are still in use. A film writer will cost upwards of quarter of a million pounds sterling, and will normally require a specially trained operator, plus the services of a skilled photographic laboratory. Few organisations using remote sensing operate their own film writer because of the high capital and running cost, but prefer to send processed tapes for film writing by a bureau service. At a significantly lower investment than for a film writer, it is possible to purchase an intermediate type of film recorder, capable of recording larger images than a simple screen camera, but not of such high resolution (or cost) as a film writer. Screen cameras can be exactly what the name implies, photographing an actual colour monitor, either the one used for interactive processing or an identical one installed in a fixed position relative to a suitable automatic camera. A more sophisticated variant is the "matrix camera", in which the output from the video monitor sequentially produces red, green and blue images on a small monochrome flat screen CRT which is photographed through a series of filters by a fixed camera system with automatic exposure control. Both of these systems only permit recording of the image displayed on the monitor, with the same restrictions on numbers of pixels and lines. Depending on the format of camera used, the resulting transparencies or colour negatives can be enlarged for later use or display. Apart from these photographic techniques for hard copy production, there are a wide range of printers and plotters which can either produce rapid colour screen dumps at various resolutions and with differing fidelity, or, in the more expensive plotters, produce full-scene full-colour plots off-line to a pre-determined scale. Even these large precision plotters are usually much cheaper than film writers, and although the product may not be as aesthetically pleasing as a large photographic print, they are adequate for most actual working purposes. Some companies offer bureau services with these large plotters, printing images and derived maps from client's own tapes.

2.5.7 Software

This is the key to the usefulness and performance of any image processing system. Even the most advanced, fastest and most powerful hardware, with the best monitors and hard copy devices, is of little real practical use unless the software can provide the whole range of required functions and operates in a way which is easily understood and learnt by the user. Innovative hardware seems sometimes to be used almost as an excuse for poor software, and it is rare to find an image processing system where state of the art hardware is taken full advantage of by fully functional and logically constructed software. One reason for this is that there should always be three different sets of specialists working together on the development of a really good image processing system. The requirements should first be defined by experienced users, preferably with experience of a range of applications of remote sensing. Hardware specialists should then build or select suitable hardware, and software should be written by another group of specialists, who should remain in constant interaction with the users. Image processing systems designed for use in a different field of image processing from remote sensing, for example medical imaging or surveillance photogrammetry, are sometimes adapted for what the manufacturers perceive to be remote sensing requirements, and are rarely successful. A system which is originally designed to process single-band eight-bit images, and is then modified to handle multi-band imagery is rarely as good as one which is designed from the start as a remote sensing system. Some of the best systems are those which have been in regular use for a period of more than five years after being designed originally for remote sensing requirements. Provided that the manufacturers maintain close contact with their customers, and that the customers know what they want, the system will become refined and more capable over the years. The disadvantage is that these older-established systems may not use state of the art hardware, and can sometimes be expensive. Software is, in my opinion, more important than the latest hardware. It is of no particular advantage to be able to do a fast Fourier transform in real time on the latest hardware, if the system will not easily produce a simple colour composite image or carry out a multi-band image classification.

2.5.8 Choosing a System

The choice confronting a prospective purchaser of image processing equipment has now become so wide as to be rather intimidating. There are at least eight manufacturers of large image processing systems specifically orientated at the remote sensing market, and at least twelve different PC based systems are available. The choice between large (mini- or supermicro-computer based) and small (PC based) systems will depend largely on the volume and type of work expected. Any organisation which plans to regularly process full satellite images, either for hard-copy production or for classification or other automated processing of large areas, will need one of the larger systems, especially if the imagery is to be geometrically corrected. Organisations interested mainly in detailed processing of relatively small areas (less than about 2000 by 2000 pixels) would probably be satisfied with a PC based system. Larger organisations requiring a number of workstations for use by geologists and other specialist staff might require a central large system networked to many smaller PC systems. The PCs can, of course, be used for standard data processing and word-processing operations as well as for image processing. The amount of remote sensing carried out in house by an organisation would have to be very large to justify the high capital and running costs of a film writer, although if bureau services were not easily available there might not be a viable alternative.

Whatever system is purchased, it should be based on widely available and well-tested hardware. State of the art equipment is not very useful if it is the only one of its kind in the country, with the nearest distributor and service centre ten thousand kilometres and two continents away. If the system is operating in a remote location, spare sets of any special boards (for example, the frame store in a PC system) should be purchased. Although the majority of tape drives now have a data density of 6250 bpi as standard, purchasers should ensure that the drive selected can operate at this density. An old-style 1600 bpi drive will make it more difficult to obtain digital data, and will triple the volume of the tape archive. Unless the organisation plans to move the image processing system often and portability is very important, the best quality colour monitors affordable should be purchased. The best monitors are often the heaviest, but good colour contrast, lack of distortion and absence of flicker are most important if the system is going to be used for hours at a time. A very large screen is not essential unless groups of people need to view the imagery simultaneously. The choice between an off-screen camera, using a standard 35 mm camera permanently mounted in the image processing room, and a separate matrix camera will be based on cost and convenience. The matrix camera will be considerably more expensive, but will free users from concern about room lighting and reflections from the surface of the monitor, as well as giving better pictures. Unless it is planned to produce very large numbers of high-quality photographic products, investment in a film writer is unlikely to be justified, although it may be wise to purchase a high-quality ink-jet or electrostatic plotter to produce working plots which can be used in the field. Most developed countries, and an increasing number of less developed ones, have film writers in national institutes or commercial organisations, and a bureau service is usually available. It is wise to check on the digital data formats required by any regional film writing service, to ensure that the combination of image processing software, operating system and tape drives that it is proposed to buy can generate suitable output tapes.

3
Geographic Information Systems In Remote Sensing

Increasing use of remote sensing in the minerals industry depends to a large extent on the integration of remotely sensed data into geographic information systems. GIS techniques are becoming accepted in the industry, and many exploration and mine managers are already using these systems without necessarily being aware of it. While some GIS applications, for example storing and analysing information about the distribution and conditions of water, electricity and similar distribution networks within a mining site, are either at an inappropriate scale or thematically unsuited for remote sensing input, there are many others to which remote sensing can make a vital contribution. Individual applications will be discussed elsewhere in the book, but some fundamental concepts benefit from being discussed first.

A critical factor governing the incorporation of remotely sensed data into a GIS is the digital nature of the data. One of the most expensive and time-consuming aspects of any GIS is data input, since much of the data to be input during the transition from analogue to digital handling of geographical data must first be digitised. Satellite imagery is already in a digital form, and can be relatively simply transformed to any geographical coordinate system or projection. The cost of transforming satellite imagery into a form acceptable to most digital GIS's is very low, and this makes its use attractive, either as a substitute for more conventional types of data, or as a totally different source of information. Digital imagery, from satellite or airborne scanner, can be used in GIS's in a more or less raw form, when its most common function is to provide an easily interpretable reference background to other data sets, or the imagery can be extensively processed and the products of this processing, improved geological maps, spectral anomaly maps, lineament maps, land-use classifications, change maps, or any others of a whole range of other derivative products, then incorporated into the GIS. The use of GIS's can also improve the processing and interpretation of remotely sensed imagery. It is rare that this imagery can yield a unique and unambiguous product without the benefit of other knowledge. The inclusion of ground data, for example lithological information, mineral occurrence and production statistics, geochemical and geophysical maps or topographical data, is often essential to efficient processing of the satellite imagery. This is conventionally done in an analogue fashion, comparing existing maps on a range of scales with photographic products of remote sensing, but subtle relationships are often not apparent unless the different information sets are actually co-registered with each other, and the analysis of inter-relationships of more than a few sets of data of totally different types is beyond the capacity of the human eye and brain and requires the speed and capacity of a computer. The concept of synergy, where the whole can often be significantly greater than the sum of its parts, has already been discussed, but should be emphasised again here. In mineral exploration particularly, associations and relationships between apparently unrelated data sets can often emerge when the data are co-registered in a GIS and these relationships can point to significant new exploration targets. Similarly, different remote sensing systems, looking at the earth at different wavelengths, resolutions or times, can in combination provide better or different information than any sensor used alone. There is a danger, however, in taking the concept of synergy too far. The combination of ever greater numbers of disparate data sets will not necessarily always lead to better understanding, and the investigator should always have a clear idea of what is being sought, and of the theoretical basis of associations discovered. The blind combination of huge amounts of data in the hope that something new will turn up is unlikely to be profitable, and may only result in clogging the computer system and preventing other more essential work being done.

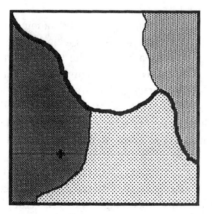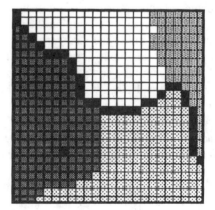

Figure 3.1. A comparison of vector (left) and raster (right) maps.

GIS's can operate in either vector or raster domains, or sometimes in combinations of both. Vector and raster formats are illustrated diagrammatically in Figure 3.1. Vector systems store line, point and area information as a series of coordinates and vectors. The thickness of line features, or the size of point features or symbols, is essentially independent of the scale at which the information is displayed or plotted, and since only the information relating to lines and other features need be stored and processed, and not the information on blank spaces between the lines, the amount of memory required to store even very complex maps is not excessive. Raster systems store and process information as raster images. Each pixel has a value assigned to it, and lines, points and areas are defined by arrangements of square pixels. The resolution and clarity of lines and other features is very dependent on display or plot scale, since a line will be a fixed number of pixels wide. Since every pixel of the image must be stored, even if the value of the pixel is zero, the amount of memory required is usually much greater than in a vector system, although it is governed by the map size, and not by its complexity. Editing of features is usually easier on a vector system, since nodes and vertices of lines can be moved, and it is not necessary to erase and re-draw every element of that line, as is necessary with rasters.

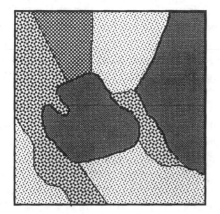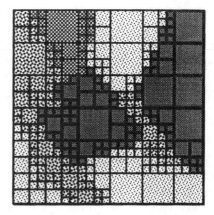

Figure 3.2. A diagrammatic illustration of the way in which a vector map (left) can be stored as quadtrees (right).

A special type of data storage and processing system, which is essentially a raster system, but modified so as to have many of the benefits of vectors, is the quadtree. This is very effective in storing information on areas, rather than lines or points, and is often used in combination with vector presentation of lines. The principle of quadtrees is illustrated in Figure 3.2. In a quadtree system, a map or image is broken down into a series of squares, each one of uniform attributes. A uniform area of complex shape is stored as a series of squares of different sizes, each smaller square being one quarter the area of the next larger size. Large uniform areas are stored as large squares, while detailed boundary areas may require large numbers of much smaller squares. A level 1 quad can be the size of the whole study area, level 2 a quarter of that, level 3 one sixteenth, and so on. In all but the largest or most complex of maps, ten or eleven levels of quadtrees are sufficient to store all the required detail. The quadtree system has many of the advantages of raster systems, especially the capability to compare attributes of areas from different, and often multiple, overlaid maps, but requires only a fraction the storage space of a standard raster system. It is not as well suited to handling line or point data as a vector system.

The ideal remote sensing GIS would combine rasters and vectors in a such a way that the user would be unaware that different storage and processing systems were being used. Remotely sensed imagery must be handled in a raster format, since each pixel has its own unique characteristics, especially in multispectral images. Once digital imagery has been processed on a raster-based image processing system, the results of this processing, which will usually be line or area maps, can be converted to vectors and processed along with other vector maps. Conversely, the digitising, editing and processing of maps is most efficiently carried out in vector format, although derivative maps may be converted to rasters for processing with satellite imagery on an image processing system. Rapid conversion between raster and vector formats is still rare. It is not too difficult to rasterise a vector map, in order to produce an image that retains at least some of the attributes of the original vector map, either in the form of digital values for each pixel of an eight-bit image, or in the form of a series of overlaid single-bit image planes. The reverse of this process is much more demanding, and is the subject of much current research. The actual conversion of a raster image to a crude vector file is simple, but the structuring of that vector file into continuous lines with attributes is much more complex, and considerable editing is usually required. This is not a great problem in many remote sensing operations, since the raster images to be converted to vectors are often either simple line maps with fixed attributes, for example lineaments derived from filtering a satellite image, or new roads interpreted from imagery which need to be added to existing maps, or else a series of polygons with attributes such as land cover or rock type derived from interpretation, automated or manual, of satellite imagery. The editing required to structure these relatively simple vector maps is not demanding, although any improvements in automated structuring of vector files are of course to be welcomed.

The main requirement for automated vectorisation is in the input of map information into vector-based GIS's. Although an increasing number of countries are producing new topographic, and sometimes even geological maps, in digital form, a dauntingly high proportion of the world's maps still only exist in analogue form as printed, or sometimes even hand-drawn, maps. The conversion of paper maps to a digital form is the greatest obstacle facing any would-be user of GIS's, and can often cost orders of magnitude more than the hardware and software themselves. This is a fact often ignored by purchasers of GIS equipment, and it is regrettable that manufacturers of such equipment rarely include warnings to this effect in their sales presentations. The most rapid technique for map digitisation, and also the cheapest in terms of equipment and time, is scanning. This can be carried out using a modified television camera (video digitising or scanning), or by means of a more precise digital scanner, which operates like a rather refined photocopier. The former equipment is usually subject to considerable geometric distortion, while the latter is geometrically and radiometrically better, but usually slower and more expensive. Digital scanners range in size from A4 up to A0 or larger, and in complexity from monochrome 4-bit systems to full-colour eight or even twelve-bit quantisation. Some specialised scanners actually follow individual lines on analogue maps, creating partially structured vector files directly, but the end result of most scanning operations is a raster image of the scanned map, either as a single-band monochrome image or a three-band full colour image. Vectorisation of this complex image to produce a structured vector map, in which all the attributes of the original paper map, in terms of line-type and colour, are stored in digital form, is the ultimate goal, but only the most expensive currently available GIS's can come close to this ideal, and none achieve it completely. There are solutions, although none of them are entirely satisfactory. If the master plates for the original map, or monochrome prints from them, are available, then these can be scanned separately. A contour plate, for example, could be scanned to give an image consisting only of contour lines, which would then only require addition of the height attributes for each line in order to be used in a GIS. Subsequent scanning of the roads plate at the same scale would allow the two to be overlaid within the GIS and processed together. It is often impractical, or impossible, to obtain such original plates. An alternative is for the maps to be traced manually, with each feature on a different tracing sheet. These can then be scan-digitised separately, although this process involves considerable input from a skilled draftsman. Bureau services are available in many countries to prepare digitised maps in a range of possible formats to suit most commercial GIS systems, but this is an expensive route if large numbers of maps are to be digitised. The problem of input and conversion of analogue paper maps is fortunately a transitory one. Firstly, digital maps are becoming more widely available in many countries, and secondly the hardware and software to effect this conversion is rapidly becoming faster and more affordable.

A critical feature of vector-based GIS's is the capability for simple access to relational databases. Each line, point and polygon in a map can have a whole range of attributes of varying complexity. For example, faults in a geological map might have information on their type (strike-slip, thrust etc) and their apparent age, stored in a relational database, while areas of specific rock units could have associated attributes such as age, lithology, permeability and magnetic susceptibility stored in the database. In a practical GIS, this database information could be used to control the display, in order to highlight all rock units of Devonian age, or all left-lateral faults, for example, or could provide inputs to exploration models. It should be possible to query the database from the display, pointing at an area, line or point to extract a selection of information from the database. This is extremely difficult to achieve in raster systems because of the requirement for each of thousands of pixels to have a complete database file. With geometrically corrected imagery and a vector system operating in parallel to an image processing system, a link could be established so that the raster coordinates of a query point are sent to the vector system to interrogate the database.

The hardware configuration of a computer-based GIS can vary as widely as an image processing system. The display requirements for GIS are not normally as complex as for image processing. An eight-bit display, without a large frame store, is usually perfectly adequate. Fast processing is extremely important, especially for re-writing vectors to the display after editing operations or when the area of view is changed. Few things are more frustrating, or wasteful of professional time, than to spend a total of more than fifty per cent of a working day waiting for maps to be re-drawn on the screen. The author had personal experience of this, using a very widely accepted GIS software system that had been installed on an inadequate computer. Upgrading of the computer had a dramatic effect, and reduced waiting time to less than five per cent of the total.

Many GIS software systems are now commercially available, at a wide range of prices and capabilities. Some are tied to custom-built hardware, but most are largely host-independent. It is unfortunate that exaggerated claims are sometimes made for systems which are actually little more than computer aided design packages, suggesting that these are actually full GIS's. The distinguishing features are the presence of a fully relational database, and the capability to analyse and display relationships between point, line and area data. Some GIS's have either been specially designed with remote sensing links in mind, or else have well-proven communication routes with established remote sensing image processing systems. Some cartographic systems combine raster processing of remote sensing imagery and vector processing of maps with the capability of overlaying both types of data. These systems are optimised for topographic mapping, and especially map updating. Although very impressive in many respects, the range of image processing functions is usually limited, and relational database operations are rarely possible. Users of PC-based systems, either for image processing or GIS operations, considering expansion into the parallel world should be cautious of overloading their system. Although image processing and GIS software can often operate on similar PCs, a very large storage capacity is necessary in order to store both suites of software and still permit sufficient working space for storage of vector files, and even more critically, to store raster images which are extremely demanding in terms of file size. It may often be more practical, and not really much more expensive, to purchase a second PC and to connect the two systems through a suitable network.

4

Current And Future Sensing Systems

4.1 OPERATIONAL SYSTEMS

A very wide range of remote sensing systems, both spaceborne and airborne, are available to the minerals industry, and new systems, some replacements to and modifications of current sensors, and others promising revolutionary new opportunities, are under development. Most current sensors were not specifically designed for the mining industry, and indeed many were either built as technology demonstrators, without specific applications in mind, or else were compromises to answer a wide range of requirements. Geological applications have however been more important than any other single applications group in dictating the design of some commonly used sensors. At least one of the spectral channels on the Landsat Thematic Mapper was added primarily to serve geological requirements, and most of the airborne spectrometers currently in use were built with geological uses very much in mind. One new satellite due for launch in 1992 has a sensor which is specifically designed for geological use. Some satellites are financed and designed specifically for meteorological use, but nevertheless have properties of interest to the minerals industry. Satellite sensors currently operational or planned for the first half of the 1990's are illustrated in Figure 4.1 and listed in Table 4.1.

The choice of an appropriate sensor for a given task is not always easy, and indeed it is rare for a single sensor to have characteristics which are ideal for a specific problem. Sensor design is usually a compromise between the conflicting requirements of diverse user communities, constrained by spacecraft power and payload, data transmission rates, and by the principles of optics, electronics and atmospheric physics. The main factors to be considered in the choice are the spatial and spectral resolution, the requirement for up to date imagery, and the cost. In general, the finer the spatial resolution, the greater the cost per square kilometre. This is illustrated for current imagery in Figure 1.10. If general information is required over a very large area, it may not be economically justifiable to use very fine resolution imagery. Equally, if spectral resolution is not important for a given application, single-band imagery may be preferable to multispectral data. In some cases, particularly in areas of the world subject to extensive and persistent cloud-cover, the choice of imagery may be dictated more by availability than by any special characteristics of the sensor. In some applications within the mineral industry, the most recent imagery is not essential, and it may be cost-effective to use older lower cost imagery rather than the most recent. This section presents most of the current spaceborne and airborne sensors in turn, and summarises their actual and potential usefulness to the minerals industry. New sensor systems planned for the coming decade are also discussed, and finally some recommendations are presented regarding imagery which is appropriate to each main group of application. Cost figures are relative in most cases, since actual costs vary from year to year, although where actual figures are quoted, it should be remembered that these are the costs in Europe as of the middle of 1991.

Table 4.1. Spaceborne sensor systems listed in order of spatial resolution

Sensor	Years of Operation	Notes
Meteosat	1984 - present	Meteorological geostationary satellite. Very frequent coverage. Coarse resolution.
AVHRR	1975 - present	Meteorological polar orbit. Frequent coverage. Coarse resolution. Multispectral with thermal.
MSU-SK	1988 - present	Moderately frequent coverage. Medium to coarse resolution. Multispectral with thermal.
Landsat MSS	1973 - present	Infrequent coverage. Medium resolution. Multispectral.
MOS-1	1988 - present	Infrequent coverage. Medium resolution. Multispectral.
IRS LISS-1	1988 - present	Infrequent coverage. Medium resolution. Multispectral.
MSU-E	1988 - present	Infrequent coverage. Medium to fine resolution. Multispectral.
IRS LISS-2	1988 - present	Infrequent coverage. Fine resolution. Multispectral.
Landsat TM	1983 - present	Infrequent coverage. Fine resolution. Multispectral with mid-infrared and thermal.
SPOT XS	1986 - present	Infrequent coverage. Fine resolution. Multispectral. Stereoscopic.
SPOT PAN	1986 - present	Infrequent coverage. Very fine resolution. Panchromatic. Stereoscopic.
SEASAT	1978 only	Partial global coverage. Digital SAR imagery.
SIR-A	1983 only	Partial global coverage. Optically processed SAR imagery.
ERS-1	July 1991	Partial global coverage. Optimised for oceanographic applications.

4.1.1 Meteorological Satellites

As their name implies, these satellites have been designed primarily for meteorological observations. Their use has become so widespread in weather monitoring and forecasting that the technology has become almost commonplace. Meteorological use of remote sensing has reached a state of maturity which has been achieved by no other application, and the contribution to greater economic productivity, safety, and general standard of living is often not fully appreciated.

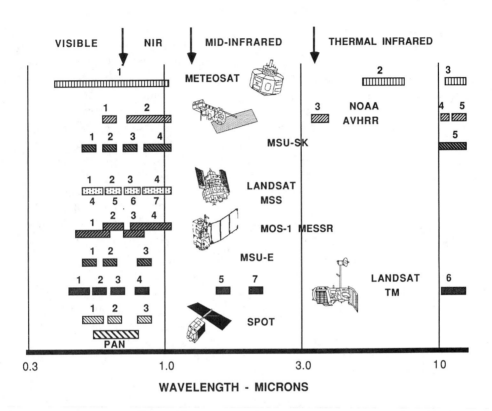

Figure 4.1. Some operational spaceborne sensors, illustrating the wavebands available. The vertical axis of the diagram is increasing pixel size (decreasing spatial resolution).

4.1.1.1 Geostationary Satellites.

Two distinct classes of meteorological satellite are operational, distinguished by their orbits, which can be either geostationary or polar. The geostationary satellites, as their name implies, are in orbits that maintain them in an apparently stationary position with respect to the surface of the earth. There is a single very specific orbit in which this geostationary position can be maintained, over the equator at an altitude of 34,000 km. This has two important implications for remote sensing. Firstly, the sensors are rather far from the surface of the earth, limiting the spatial resolution which can be achieved. Secondly, while the view of equatorial areas is close to vertical, the angle of view becomes increasingly oblique at higher latitudes, and polar regions are effectively invisible. The entire surface of the globe between latitudes of about 60 degrees north and south can be covered by four satellites equally spaced around the equator, although in practice more satellites are usually deployed to give less distorted cover. The geostationary meteorological satellite network is a good example of international cooperation, with satellites built and operated by the United States, the European Space

Agency, India and Japan. Most of these satellites have broadly similar characteristics, imaging the earth's surface in visible and thermal wavelengths, with in some cases an additional waveband designed to monitor atmospheric water vapour. Spatial resolution at nadir ranges from 10 km down to 1.5 km, and the acquisition frequency from once every thirty minutes for general images to daily for the more detailed imagery. These satellites are little used for land surface studies because of their poor spatial resolution and lack of spectral information. Their very wide and frequent coverage does however permit the observation of whole continents, and gross surface phenomena such as major floods can be monitored using these satellites. They also play a vital part in provision of early warning of extreme weather events such as hurricanes, information which is as important in the mineral industry as in any other field of human endeavour, and are used increasingly in semi-arid areas for monitoring regional rainfall. Data is transmitted from the satellites in two formats: a coarse-resolution analogue format which can be received on very simple receivers (APT or automatic picture transmission) and a finer-resolution digital format requiring a more complex receiver (PDUS or primary data user station). The data supply is currently (1991) free to all who wish to receive it, so that even remote mining and prospecting operations can obtain timely information on regional weather conditions, especially in low latitudes. Frequent use is made of such systems by organisations involved in airborne geophysics and aerial photography. Imagery from these satellites is geometrically corrected by the ground control station before re-transmission to users, and further processing is thus greatly simplified. The Meteosat satellite operated by Eumetsat on behalf of the European Space Agency is a typical satellite of this class. The sensor characteristics are summarised in Table 4.2.

Table 4.2. Characteristics of Meteosat sensors

Channel	Wavelength Microns	Resolution Km	Notes
1	0.5 - 0.9	2.5/5.0	Visible channel
2	5.7 - 7.1	5.0	Water vapour channel
3	10.5 - 12.5	5.0	Thermal channel

Despite the coarse resolution and lack of spectral information, geostationary satellite imagery can be used to produce mosaics of entire continents at a single wavelength. The imagery is geocoded to a space stereographic projection by the satellite operators, and production of a digital mosaic is thus a fairly simple task. Since the data is digital, these mosaics can be further processed to enhance major structural elements, aiding regional tectonic studies which can make an important contribution to strategic mineral exploration and area selection. An example of such a processed mosaic, prepared from 2.5 km resolution Meteosat VIS imagery (a broad-band sensor extending from the visible into the near infrared) is shown in Figure 4.2. This mosaic could make a significant contribution to the understanding of the gross geological structure of the African continent, which might have important future implications for the minerals industry.

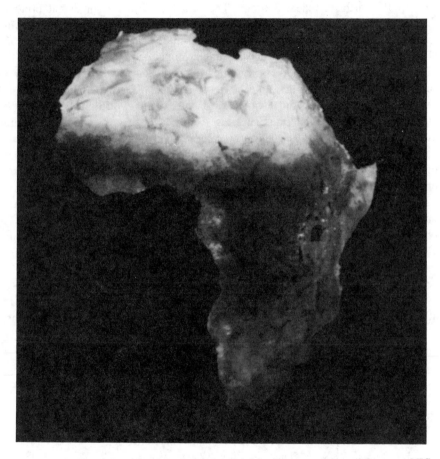

Figure 4.2. A coarse-resolution mosaic of Africa, prepared from Meteosat VIS
imagery.

Geostationary meteorological satellites are being continually improved to meet
user demands, and it is likely that the next generation, such as the second generation
Meteosat, will have greater application in the minerals industry than the earlier models.
Spatial resolution is likely to be improved to one kilometre at nadir, and near infrared
and additional thermal bands added, which apart from allowing better discrimination of
cloud types, the prime requirement for a meteorological satellite, will allow regional
vegetation studies by means of vegetation indices and permit continental-scale coarse
resolution lithological mapping.

4.1.1.2 Polar Orbiting Meteorological Satellites

Polar orbits are so named because they pass over the north and south polar regions, although not usually directly over the poles. This type of orbit uses the fact that the earth rotates in a direction at right angles to the orbit to give complete coverage of the surface of the earth on repeated orbits. The orbits are usually somewhat oblique in order to achieve the same local sun time in northern and southern hemispheres during descending orbits. The ascending portion of the orbit is usually on the night side of the earth. Polar orbiting meteorological satellites are operated by the United States, the Soviet Union and China. All these satellites transmit coarse-resolution imagery in a form that can be received by simple low-cost receivers, known usually as the APT (automatic picture transmission) format. Most parts of the world are covered at least four times daily by such satellites, and the APT imagery is more detailed at higher latitudes that than from the geostationary satellites. APT imagery can be used in the mineral industry as a source of weather information, but for any other purpose the poor spatial resolution and analogue form of the data is a severe restriction. Digital, finer-resolution data would be preferable, and is currently only widely available from the AVHRR sensor on the TIROS-N satellites operated by the American National Oceanographic and Atmospheric Administration, although the Chinese FY-1B satellite provides similar imagery for restricted areas of the world. The ATSR instrument on ERS-1 provides similar thermal imagery, although it lacks the visible and near infrared information of AVHRR.

The AVHRR/2 sensor, currently (mid-1991) operating on the NOAA 9 and NOAA 10 polar orbiting satellites, has a minimum pixel size at nadir of 1.1 km, a swath width of more than 2000 km, and observes the surface at five wavelengths, as shown in Table 4.3 .

Table 4.3. Characteristics of AVHRR

Channel	Wavelength Micrometres	Notes
1	0.58 - 0.68	Visible. Surface features.
2	0.73 - 1.10	Near infrared. Water and vegetation
3	3.55 - 3.93	Thermal. Fires, volcanoes etc.
4	10.5 - 11.3	Thermal. Lithology with band 5.
5	11.5 - 12.5	Thermal.

Channel 1 approximates to visible red, while channel 2 is in the near infrared, approximating to MSS band 7 or TM band 4. Channel 3 is unusual, this wavelength being partly reflected solar radiation and partly emitted thermal energy, and is especially sensitive to small-area high-intensity heat sources. Channels 4 and 5 are both in the thermal infrared region, approximating to TIMS channels 5 and 6 respectively. The NOAA satellites are primarily operated for meteorological purposes, and the visible and near infrared channels are intended mainly for cloud identification and classification. The three thermal channels are designed to permit accurate atmospheric correction of cloud-top and ocean-surface temperatures, and also to assist in cloud characterisation. Much use has been made of channels 1 and 2 in vegetation studies, particularly in semi-arid environments where vegetation indices derived from near infrared/red combinations can correlate closely with crop and rangeland productivity. The high frequency of repeat coverage with AVHRR (each satellite images every part of the earth's surface twice in 24 hours) makes this sensor suitable for monitoring natural hazards such as volcanoes and forest fires, and also for mapping transitory surface features such as snow cover. The possibility of acquisition of thermal imagery in the day and night of the same 24 hour period has permitted studies of thermal inertia of rocks as an aid to geological mapping and mineral exploration. AVHRR imagery has also been used to prepare mosaics of large areas subject to frequent cloud cover, such as a recent mosaic of the entire Antarctic continent prepared at the NRSC.

A common requirement in primary exploration, both for hydrocarbons and metallic minerals, is to obtain synoptic views of large areas, for example an entire country or the whole of a particular geological province. Landsat imagery has been used in exploration for almost a decade and a half, and mosaics can provide such an overview. The cost of mosaicking large numbers of Landsat scenes is however very high, and unless done digitally, which is demanding of computer space and time, the resultant mosaic cannot be further digitally processed to enhance structural features and lithology. The AVHRR sensor, with its very broad swath, can provide this overview at a fraction of the cost of acquiring and mosaicking Landsat or SPOT data. The fact that the AVHRR acquires data at three thermal wavelengths means that it can also, under arid conditions, provide thermal spectral information. The high temporal frequency of coverage allows studies of thermal inertia by comparison of co-registered day and night imagery.

Discrimination of lithological units in AVHRR imagery depends on differences in thermal emissivity with wavelength. Thermal emissivity in channel 5 (11.5 to 12.5 microns) is higher than that in channel 4 (10.5 to 11.3 microns) for most common rocks, but the ratio of 5/4 radiances is greatest for silica-rich rocks and minerals and lowest for carbonates, argillaceous materials having intermediate values (AVHRR digital values in the thermal channels actually decrease with increasing radiance, so that ratios based on digital numbers will be the converse of radiance ratios). If a ratio between bands 5 and 4 is calculated, subtle differences appear which can often be closely correlated with lithology, as shown in Figure 4.3. A crude lithological map, discriminating areas of siliceous rocks from areas with mainly carbonates and clay minerals, can thus be constructed from the thermal imagery. Intensity-hue-saturation colour composite images, combining the 5/4 ratio (hue) with a channel 2 image (intensity) and a 2/1 ratio (saturation) form readily interpretable geologically enhanced products. Plate 9 shows a composite image of this type prepared for an area in southern Morocco. The close correspondence between this image and the geological map is clear. A major limitation on the use of this technique is the effect of atmospheric absorption on the satellite-perceived radiance in channels 4 and 5. Water vapour absorption is greater in channel 5 than in channel 4, a factor which is used in correction of cloud and sea-surface temperature values to allow for water vapour absorption. The effect of atmospheric water vapour would be to change the 5/4 ratio, masking more subtle variations due to lithological differences. The thermal ratio technique can thus only be used with confidence for lithological discrimination in areas of low, or at least constant, atmospheric water vapour content, which normally means arid or semi-arid climates.

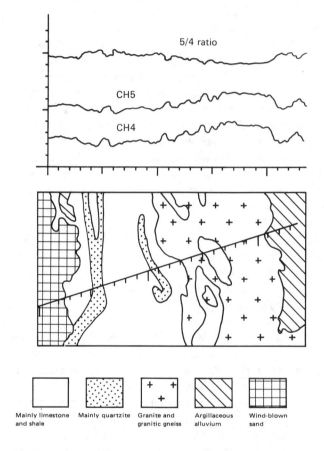

Figure 4.3. Profiles of digital values in bands 4 and 5 of AVHRR, and the 5/4 ratio, over an area of diverse geology in southern Morocco.

4.1.2 Landsat MSS

Despite its rather coarse spatial resolution and limited spectral coverage, this is probably still the most used sensor in the minerals industry. From the launch of Landsat 1 (ERTS-1) in 1972 until Landsat 4, with its Thematic Mapper sensor, was launched in 1983, the MSS was the only source of digital satellite data available to the minerals industry. Very large archives of digital and photographic products were built up during this period. MSS is a valuable historic data source for change studies, and as a low-cost alternative to TM or SPOT for regional land-cover and geological studies.

Table 4.4. Characteristics of Landsat MultiSpectral Scanner (MSS)

Channel	Wavelength	Notes
4 (1 from Landsat 4)	0.50 - 0.60	Visible green. Useful for water studies
5 (2 from Landsat 4)	0.60 - 0.70	Visible red. For NDVI, water, soils.
6 (3 from Landsat 4)	0.70 - 0.80	Red to near infrared. Little used
7 (4 from Landsat 4)	0.80 - 1.10	Near infrared. Important for NDVI, topographical mapping

An example of an MSS false-colour composite of a type widely used for logistical and reconnaissance geological interpretation purposes in geological mapping and mineral exploration is shown in Plate 1. The poor spectral resolution of MSS, particularly the lack of information from the mid-infrared portion of the spectrum, restrict its use for lithological mapping. The fact that iron-stained alteration zones associated with many mineral deposits have a characteristic red colour, with strong absorption at other wavelengths, has been used as a basis for many complex ratios designed to enhance iron-bearing alteration zones.

4.1.3 Landsat TM

The TM has adequate spatial resolution for the majority of application requirements, combined with a useful range of spectral channels in the visible and near, mid- and thermal infrared. The wavebands available on the TM sensor offer greater possibilities than Landsat MSS and SPOT XS bands for vegetation discrimination. The mid-infrared TM bands, in particular band 5, have been found to be particularly useful for discriminating between different types of woodland, cereals, beet, potatoes and grassland. This mid-infrared channel is not available on either Landsat MSS or SPOT XS and this represents a major drawback for vegetation studies using these sensors.

Table 4.5. Characteristics of Landsat Thematic Mapper

Channel	Wavelength	Notes
1	0.45 - 0.52	Visible blue, water and urban studies
2	0.52 - 0.60	Visible green, water and urban studies
3	0.63 - 0.69	Visible red, water studies, vegetation
4	0.76 - 0.90	Near infrared, vegetation and topography
5	1.55 - 1.75	Mid-infrared, vegetation, especially forestry
6	10.4 - 12.5	Thermal (120 metre resolution)
7	2.08 - 2.35	Mid-infrared, mainly for soils and minerals

The 30 metre spatial resolution of TM is appropriate to many land-cover classification applications, since it results in an averaging effect reducing the heterogeneity of many cover types while allowing sufficiently accurate definition of land parcels (fields, woods, urban areas) and linear features such as roads and rivers. Finer resolution is however required for applications such as monitoring activities within mineral working sites, and TM imagery on its own cannot usefully be enlarged to scales of greater than 1:25,000. Combinations of TM and SPOT PAN imagery are particularly useful, since they combine the fine spatial resolution of the PAN with the spectral information of TM.

In oceanographic studies, TM is the only operational sensor that can be used for ocean colour work, since the visible channels are similar to those of CZCS, and the fine spatial resolution allows investigation of small scale features. The disadvantages are that the area covered by a single image is small compared to many of the oceanographic features of interest, and that the TM spectral channels were not specifically designed with oceanographic work in mind. The band width of visible channels is too wide for discriminating between different biological pigments in the water. Sensor gain over water is not sufficiently high to produce a satisfactory signal to noise ratio. The sensitivity of the thermal band is too low for many oceanographic applications, although its value as the finest spatial resolution spaceborne thermal sensor currently available has been demonstrated in projects mapping, for example, distribution of thermal plumes in power station cooling waters. Calibration of the thermal data has to be carried out for each image using surface data due to the drifting of the on-board gain values.

The lithological discrimination potential of TM, and in particular of bands 5 and 7, is extremely important in geological studies. Band 7 is in a spectral region where many clay minerals show strong absorption peaks, and can be used in combination with other bands to highlight areas rich in clay, such as hydrothermal alteration zones. Band 5, when used in combination with visible imagery (usually band 3) can highlight areas rich in iron, and so can assist in the location of gossans and other areas of more subtle iron enrichment. These uses are described in greater detail in a later chapter. Of all the currently available systems the TM is the optimum for structural analysis, as its spatial resolution allows structural filtering enhancement at both local and regional scales.

4.1.4 SPOT HRV and PAN

Table 4.6. Characteristics of the HRV sensor on SPOT 1 and 2

	Multispectral Mode (XS)	Panchromatic Mode (PAN)
Spectral Bands	0.50 - 0.59	0.51 - 0.73
	0.61 - 0.68	
	0.79 - 0.89	
Pixel Size	20 metres	10 metres
Swath Width	60 km	60 km

The French SPOT series of satellites, first launched in 1986, provide the finest spatial resolution digital satellite imagery currently commercially available, and have significant applications in the mineral industries. SPOT 1 and SPOT 2 satellites each carry two pushbroom scanner systems, one operating on each side of the spacecraft track, and each capable of acquiring imagery in two different modes, multispectral and panchromatic. Unlike the Landsat satellites, the SPOT sensors are steerable by means of tilting mirrors. This has two important consequences: firstly that acquisition of imagery is not rigidly tied to a timetable based on the orbit repeat cycle. The steerable sensors can be used to acquire off-nadir imagery from adjacent nominal image paths, allowing a much more flexible acquisition schedule than with the Landsat satellites. The second major consequence of the steerable sensors is that the same area can be viewed at two or more different look angles, permitting acquisition of stereoscopic imagery. The fact that the stereo pairs are not acquired within a very short time interval of each other, but from different orbits separated in time by days or even weeks or months if cloud cover is severe, is a problem where surface features such as vegetation change rapidly, and will be rectified in future spacecraft by the provision of a mirror system which tilts along-track rather than across-track. The sensor characteristics are listed in Table 4.6.

The fine spatial resolution and stereoscopic nature of SPOT imagery, particularly the PAN mode, make it ideally suited for topographic mapping applications. Many remote areas of the world, where reliable topographic maps did not previously exist, have been mapped with the aid of SPOT PAN imagery, and a number of software suites for automated extraction of digital elevation models and contour maps from stereo imagery have been developed commercially. The stereoscopic capability is also of importance in structural geological studies, and imagery can be treated like a small scale air photo for visual interpretation using a stereoscope. The flexible repeat coverage offered by the steerable sensors increases the chances of acquiring imagery of cloudy areas, for example in equatorial regions, and also permits rapid observation of remote areas in times of emergency. Studies of rapid temporal change associated with natural disasters can be carried out if programming of repeated imagery can be arranged.

There are two main obstacles to the greater use of SPOT imagery in the minerals industry. One is the lack of spectral information compared to Landsat TM. SPOT multispectral (XS) imagery has only three wavebands, two in the visible portion of the spectrum and one in the near infrared, roughly comparable with TM band 4 or Landsat MSS band 7 (4 in recent satellites). Landsat TM has two additional bands in the mid-infrared, one at 1.5 microns which provides critically important information for vegetation mapping, and the other at 2.2 microns which is especially sensitive to mineralogical variations in surface soils and rocks. SPOT imagery does not permit as much discrimination of surface types as is possible with TM. It is difficult to define alteration haloes with any confidence, for example, or to distinguish lithological units based on their clay mineral, silica or iron contents. The second major problem, particularly in regional exploration programmes, is the relatively high cost of SPOT imagery per unit area compared with Landsat. If the area of interest is less than that covered by a full SPOT scene (60 km square), then the high cost is unlikely to be a limiting factor, but for an organisation interested in mapping the geology of an entire country the size of, for example, Iran, the cost difference would be very significant, particularly when the information content for geological purposes is likely to be considerably lower.

The main application of SPOT imagery in the mineral industries would appear to be in logistical planning. The imagery permits map-making and map-updating to scales of up to 1:20,000, and the stereoscopic capability allows automated generation of digital elevation models and contour maps. There is also the possibility of combining SPOT PAN imagery with a more informative multispectral data source, particularly Landsat TM, to produce composite images with the spectral information of TM and the effective spatial resolution of SPOT PAN. SPOTIMAGE, the operators of the SPOT satellites, now prepare and market a digital product that combines the multispectral information of SPOT XS imagery with the spatial resolution of SPOT PAN. This imagery, known as PXS, still has the spectral limitations, from a geological point of view, of SPOT XS, but for an organisation without its own digital processing capabilities could be an extremely useful logistical tool.

4.1.5 IRS-1

India is one of the largest users of remote sensing data in the world, and has developed its own remote sensing capability as well as making use of other nations' satellites. The Indian remote sensing satellites IRS-1A and IRS-1B are designed mainly for agricultural and land use studies, and do not provide global coverage. The primary objective is to provide imagery of the Indian subcontinent, and routine acquisition is only within the reception range of the main Indian satellite ground station in Hyderabad. This means that coverage of India, Pakistan, Sri Lanka, Afghanistan, Burma, Thailand, the Himalayan region, and parts of the southern ex-Soviet states, south-western China, eastern Iran, and much of Malaysia is obtained on a 11 day repeat cycle (two satellites), but other parts of the world are not imaged. The two operational (as of late 1991) satellites each carry pushbroom scanner systems, named the Linear Imaging Self-Scanning Sensors (LISS), imaging at two spatial resolutions, approximately 80 metres and 40 metres, as indicated in Table 4.7

Table 4.7. Characteristics of IRS-1 imaging systems

	LISS-1	LISS-2
Spatial Resolution	72.5 m	36.25 m
Spectral Bands	0.45 - 0.52	0.45 - 0.52
	0.52 - 0.59	0.52 - 0.59
	0.62 - 0.68	0.62 - 0.68
	0.77 - 0.86	0.77 - 0.86
Scene Size	148*148 km	74*74 km
Repeat Frequency	22 days	22 days

The wavebands are almost identical to those on the Landsat MSS, and were selected for their general utility in a wide range of applications and so as to provide continuity of data type to customers used to working with MSS data. The lack of mid-infrared information comparable to that obtained in bands 5 and 7 of Landsat TM is a problem for vegetation mapping and some geological studies, but was dictated by the problems of operating CCD arrays in this portion of the spectrum, as is the case with the first generation of SPOT satellites.

Any mining organisation working within the area of coverage of IRS-1 would be well advised to investigate this source of data. The costs per unit area are low compared with Landsat and SPOT, and for some logistical and general vegetation mapping purposes, as well as for geological structural studies, IRS-1 imagery, especially that from the finer resolution LISS-II sensors, could serve as a stand-alone data source. In other applications, the relatively low data cost enables its use in multi-date studies to monitor change or gather information on seasonal variations, and increases the chances of cloud-free image acquisition in cloudy areas since the imaging dates will rarely coincide with Landsat or SPOT.

4.1.6 Digital Optical Imagery from the former Soviet Union

At the time of writing (late 1991) the only digital imagery acquired at optical wavelengths and marketed in the "West" by Glavkosmos and other sections of the former Soviet space programme is that from the Resours-O sensors on the Kosmos 1939 satellite. The characteristics of these sensors are listed in Table 4.8.

Table 4.8. Characteristics of "RESOURS-O" Sensors

	MSU-E	MSU-SK
Swath width, km	45	600
Pixel size at nadir, metres	45	170
		(600 for band 5)
Sensor wavebands, microns	0.5 - 0.6	0.5 - 0.6
	0.6 - 0.7	0.6 - 0.7
	0.8 - 0.9	0.7 - 0.8
		0.8 - 1.1
		10.4 - 12.6

MSU-SK is apparently designed for regional monitoring purposes, with a broad swath, medium spatial resolution, and five channels of spectral data. The wavebands of the MSU-SK sensor are apparently identical to Landsat MSS, with the addition of a 600 metre resolution thermal channel. A single thermal channel would be subject to the same calibration and correction problems as Landsat TM band 6, and could not be used for quantitative thermal remote sensing. The finer-resolution MSU-E sensors, one nadir-pointing and the other steerable up to 350 km either side of the nadir point, permit selective acquisition of more detailed information within the broad swath of the MSU-SK. The MSU-E wavebands correspond approximately to Landsat TM bands 2, 3 and 4 and to the three bands of SPOT XS. Since the satellite is in a sun-synchronous polar orbit, at an orbital altitude of between 630 and 830 km, the repeat cycle for the broad swath coverage would be five days with 17% overlap. Total coverage with the finer-resolution sensor would be acquired once in 56 days, although other programming constraints would make complete coverage unlikely. It appears that total repeat coverage with the MSU-E was never intended, but that the normal operational

mode is to acquire repetitive total coverage with the MSU-SK, using the
finer-resolution MSU-E to acquire selected more detailed imagery as required. An
example of a near infrared MSU-E image of open-cast mines in Ukraine is shown in
Figure 4.4.

Figure 4.4. Part of an MSU-E Band 3 (NIR) image showing open-cast mining
operations in Ukraine (courtesy Glavkosmos).

The image examples available suggest that, while data quality in the near
infrared (band 3 for MSU-E and band 4 for MSU-SK) is good, with a large dynamic
range and relatively low noise levels, the same is not true for bands 1 and 2 of the
MSU-E (see Figure 2.4). The visible wavelength images show pronounced vertical
striping characteristic of "pushbroom" sensors, accentuated by the small dynamic

range. A low signal to noise ratio results in random speckle which is particularly prominent in band ratio images. Studies of inter-band correlation indicate that, as would be expected, and as is also characteristic of SPOT, channels 1 and 2 are quite strongly correlated.

The poor data quality and relatively coarse resolution of MSU-E imagery makes it unlikely that this would be much used in the minerals industry in areas where any other kind of digital satellite imagery was available. MSU-SK imagery has a spatial resolution which fills a very important gap between coarse resolution meteorological satellite imagery and the finer resolution of most earth-observation satellites. MSU-SK could be very useful for vegetation monitoring purposes if the data was easily available on a regular basis, and might also be of value in continental structural geological studies.

4.1.7 Soyuzkarta Photography

The former Soviet Union, although lagging behind the West in commercial scanning sensors for remote sensing, has made much greater use of spaceborne photography. The recovery of film from satellites has become highly developed, and the very large number and relatively short life of Soviet satellites makes spaceborne photography, with its requirement for large volumes of consumables, and for physical return of the imaging medium, exposed film, without undue delay, a more practical proposition than it would be in the West. A wide range of different camera systems are available, and new ones are continually being developed and tested, but broadly speaking they can be classified into medium-resolution multispectral cameras and fine-resolution systems with very limited spectral capability. The multispectral cameras have up to six wavebands, and thus six separate cameras fitted with narrow-band filters. The visible and near infrared portions of the spectrum are normally covered, and colour composite and other derivative images are produced by combining negatives acquired at different wavelengths. The equivalent spatial resolution of these systems is probably not better than forty metres, and the area of coverage can be large. The fine-resolution systems, of which the KFA-1000 is the best-known example, are single-camera systems using either a panchromatic film or a special two-layer emulsion, with one layer sensitive in the near infrared and the other in visible light. This results in a type of false-colour infrared picture. The equivalent spatial resolution of the KFA-1000 is between five and seven metres, although the area of coverage is less than for the shorter focal length systems.

Soviet space photography has been little used in the West for a variety of reasons, the first being that it is only since 1987 that these products have begun to be marketed in the West. A second problem is that it is difficult for prospective customers to obtain information about archive holdings, and that delivery of data is very slow. The major factors which limit its use are probably technical rather than commercial. The analogue nature of the photographs limits the amount of processing which can be carried out. It is possible to digitise the imagery, but this is expensive and results in loss of quality. The multispectral imagery lacks vital information from the mid-infrared, and information extraction is limited by the need for optical rather than digital processing. The main advantages of the fine-resolution KFA-1000 photography are its spatial resolution, which is superior to any other current commercial spaceborne sensors, and the possibility of stereo coverage. The imagery could be used for cartographic purposes in developing countries, or even for map-updating in the better-mapped countries, although processing would be by conventional analogue techniques, and this belongs more to the field of photogrammetry than to remote sensing as it is normally thought of in the West.

For economic reasons, and because of a scarcity of digital image processing hardware, extensive use of space photography has been made in the former Soviet Union itself, as well as in the countries of Eastern Europe. Some mineral exploration projects in developing countries, carried out by organisations from what was known as the Eastern Bloc, have also made extensive use of space photography, although the cost of this imagery in the West is equivalent to that of SPOT digital data.

4.1.8 MOS-1

Table 4.9. Characteristics of MOS-1 sensors

Channel	MESSR wavelength	Notes	VTIR wavelength	Notes
1	0.51 - 0.59	vis.green	0.50 - 0.70	visible panchrom.
2	0.61 - 0.69	vis.red	6.0 - 7.0	water vapour
3	0.72 - 0.80	red/NIR	10.5 - 11.5	thermal
4	0.80 - 1.10	NIR	11.5 - 12.5	thermal
Pixel size	50 metres		3 kilometres	
Swath width	185 kilometres		1500 kilometres	

The Japanese Marine Observation Satellite (MOS-1) carries two imaging sensors, the MESSR and VTIR, neither of which have great potential in the minerals industry. The MESSR has identical wavebands to Landsat MSS, and a similar spatial resolution, and so this imagery could be used as a substitute for MSS in a few applications. The lack of global coverage, and the low dynamic range of the data, restrict its use over land. Landsat MSS imagery, like TM and SPOT, has an eight-bit dynamic range, which means that digital values in raw imagery can range from 0 up to 255. MOS-1 MESSR imagery is quantised to only six bits, which restricts the possible range of digital numbers to 0 to 63. The combination of small dynamic range with the fact that the sensor is a "pushbroom" type, with relatively high noise levels, means that the imagery, especially in the visible bands, is very noisy. MOS-1 is not, however, designed as a land observation satellite, but is primarily intended for improved observations over sea. It has a high-gain mode, in which the low reflectivity over water, which leads to a very small dynamic range with conventional satellites, is increased to fill a significant part of the possible range of 64 digital numbers. This has considerable potential in the oil industry for detection and monitoring of oil slicks, which have a very low contrast in conventional imagery. There may also be potential in studying coastal pollution related to on-shore mineral extraction operations, although the restricted range of imagery available at the time of writing has not permitted testing of this possibility.

4.1.9 JERS-1

The second Japanese remote sensing satellite, JERS-1, is designed to be of special interest to the mineral exploration community, although its imagery will also be of use in other fields. At the time of writing it has still to be launched, but the satellite is programmed to be in orbit before this book goes to press. JERS-1 carries two main remote sensing instruments, an L-band SAR with a nominal ground resolution of 18 metres and the OPS optical sensor which combines three SPOT-like visible and near infrared bands, a near infrared along-track stereoscopic capability, and four wavebands in the mid-infrared, three of them in the geologically important 2.2 micron region. The sensor characteristics are summarised in Table 4.10.

Table 4.10. Characteristics of JERS-1 imaging sensors

	SAR	OPS
Ground Resolution	18 m	18 m
Wavelength	L-band	0.52 - 0.60
		0.63 - 0.69
		0.76 - 0.86
		0.76 - 0.86 (stereo)
		1.60 - 1.71
		2.01 - 2.12
		2.13 - 2.25
		2.27 - 2.40
Swath Width	75 km	75 km
Incidence Angle	35 degrees	15.3 deg (stereo)
Repeat Frequency	44 days	44 days

A high-capacity on-board recorder will permit recording of up to 20 minutes of SAR or OPS imagery, providing coverage in areas without ground receiving stations for the first time with a digital SAR system.

This satellite should provide a whole range of extremely useful data to the mineral exploration community. The visible and NIR bands will permit logistical mapping in more detail than current multispectral SPOT imagery, and the along-track stereoscopic capability is a significant improvement on the side-looking, and therefore non-simultaneous, stereo imagery provided by SPOT. The automated extraction of digital elevation models, and even visual interpretation, is greatly simplified if there are no temporal differences (cloud cover, vegetation changes) between the two images of a stereo pair. The 1.6 micron band will perform the same function as band 5 of Landsat TM, allowing greatly improved vegetation mapping. The multi-band capability in the 2.2 micron portion of the spectrum should permit the acquisition of detailed mineralogical information, especially in arid areas, and will thus improve geological mapping capabilities and the detection of alteration zones. The two possible problems with JERS-1, which only the passage of time after launch will indicate to be serious or not, are the quantisation level and data accessibility. For economy in data storage and transmission, the eight bands of OPS data are quantised to 6 bits (64 grey levels) compared to the 8 bits (256 grey levels) used for the SPOT and Landsat satellites. Experience with MOS-1 data, also quantised to 6 bits, has suggested that signal to noise ratios are low, and the very narrow bandwidth of the 2.2 micron channels may result in rather high noise levels, which might have been reduced if a larger dynamic range was permitted. The concerns about data accessibility originate in official statements by some Japanese agencies that the national space programme is designed primarily to assist Japanese industry. Although open access to JERS-1 data has been promised, experience with MOS-1 has not been encouraging.

4.1.10 Previous Spaceborne Microwave Sensors, SEASAT, SIR-A, SIR-B

These three relatively short duration missions during the 1970's and early 1980's served to test the possibilities of microwave remote sensing, and provided the first, and until very recently only, images from space in the microwave portion of the spectrum. For geological studies, particularly in structural geology, the time of acquisition of the imagery is of little consequence, and so the imagery acquired on these experimental missions is still used occasionally for geological purposes. The main limitation is that neither of these provided a truly global data set. SEASAT had no on-board recording system for SAR data, owing to the very large data volumes (the same problem persists with the current ERS-1 satellite), and coverage was restricted to the reception area of the limited number of ground stations which participated in the Seasat programme. Coverage of usable imagery was also restricted by the logistical problems of processing large volumes of SAR data with the computer systems of the day, and much raw SAR data remains in archive without ever having been converted to image format. Both the Shuttle missions (SIR-A and SIR-B) acquired coverage restricted to the shuttle orbit, approximately 45 degrees north and south of the equator, as well as by the short duration of the Shuttle flight. In the case of SIR-B, technical problems with the very complex SAR instrument further restricted the actual coverage that was achieved. The characteristics of these three systems are listed in Table 4.11.

Table 4.11. Characteristics of Seasat, SIR-A and SIR-B SAR imagery.

Sensor	Waveband	Incidence angle deg	Pixel Size (m)	Swath Width (km)	Polarisation
Seasat	L-band (23.5 cm)	20 - 26	25	100	HH
SIR-A	L-band (23.5 cm)	47 - 53	40	50	HH
SIR-B	L-band (23.5 cm)	15 - 60	25	variable	HH

Like all single-waveband, single polarisation SAR imagery, the imagery from SEASAT and SIR-A has rather limited use in mapping surface cover types, although experienced interpreters can extract useful information by visual interpretation of differences in surface roughness. SIR-B imagery was acquired at different look-angles, but only over very restricted areas. This imagery served to test the possibilities of future more complex SAR systems, and whetted the appetite of the exploration industry, but does not constitute an archive which will be often consulted for practical use. Figure 4.5 is a curiosity prepared from SIR-A imagery to demonstrate possible stereoscopic uses of spaceborne SAR imagery. The Shuttle orbit is such that successive tracks sometimes cross at angles of up to 35 degrees, permitting views of the same site from slightly different directions. A search of the SIR-A archive revealed that imagery was acquired for three such crossing points over land during the SIR-A mission. Two of these points were in extremely rugged terrain, one in the northern Andes and the other in Borneo, but one occurred over the island of Keffalinia, off the coast of Greece. This island is mountainous, but not excessively so, and being an island is surrounded by a plane surface of known elevation. It was therefore possible to scan-digitise the two optically processed images of Keffalinia, and to co-register one digital image to the other using control points around the shoreline. The images thus had no relative displacement at sea level, and any relative displacement away from and above sea level would result mainly from topographic elevation. A stereo pair thus resulted, which can be viewed with a pocket stereoscope. The stereoscopic effect is not perfect, since it results from different look directions and effective incidence angles rather than from a simple horizontal separation of viewpoints, but the effect of relief is quite impressive.

Figure 4.5 A stereo pair of the island of Keffalinia, off the south coast of Greece, produced from two overlapping SIR-A orbits.

Airborne radar imagery is widely used for geological studies in the humid tropics, where cloud and tree cover render acquisition of conventional air photography difficult. No SEASAT imagery of these portions of the world was acquired owing to the lack of receiving stations in the equatorial zone, but much of the SIR-A coverage was in this area, and some of it has been used in exploration programmes. The low incidence angle of most SAR imagery compared with solar illumination angles over much of the world as observed by the passive optical satellites results in enhancement of surface relief in SAR imagery which can assist in structural interpretation. There are two serious problems associated with this relief enhancement, however. One is that it has a strong directional bias. Features sub-parallel to the satellite track are strikingly enhanced, while those at right angles to the track are almost invisible. The second problem is that relief produces serious distortions in SAR imagery. In very mountainous areas these distortions can be so extreme as to render the imagery unrecognisable, but even in more subdued topography the distortions seriously degrade the geometric fidelity of the imagery, and restrict its usefulness for topographic mapping purposes. Geometric correction of the imagery is a very computer-intensive and therefore expensive process, which must rely, in the case of current spaceborne microwave sensors, on extrinsic information from topographic maps and digital elevation models. This renders correction impossible in just those areas where it is most needed, where accurate topographic maps do not exist. Some airborne SAR systems acquire imagery simultaneously at different look angles, permitting stereoscopic processing, extraction of digital elevation models and automated correction of the imagery, and this is likely to be feature of future more advanced spaceborne SAR systems. An additional problem with SAR imagery which restricts its use in mapping and logistics is the high noise level. Semi-random noise, or "speckle" is characteristic of all SAR imagery, and is in part an artifact of processing. Its effect is to greatly reduce the effective spatial resolution of the imagery, except for features having a very strong backscatter contrast with their surroundings. Spatial resolution of SAR imagery cannot be directly compared with that obtained with optical sensors, and for most purposes SAR imagery processed to a pixel size of 20 metres can best be compared with optical imagery of 60 metre pixel size.

Although these limitations are severe, imagery from SEASAT and SIR-A has been used in some minerals applications, and will still continue to be consulted even when new imagery is available from current and planned microwave satellites. Particularly when used in combination with imagery from optical sensors, this SAR imagery can provide much useful structural information, and even additional lithological information in vegetated terrains. The fact that L-band microwave energy can penetrate a significant thickness of dry sand has permitted structural studies of some sand-covered desert areas impossible with optical sensors. Since these were experimental, non-commercial missions, imagery is relatively cheap. SIR-A imagery was optically processed, and so is not available in digital form, but film products permit an extremely low-cost overview of some remote tropical regions. It is possible to scan-digitise the film products in order to allow geometric correction and co-registration with other data sets. The main use of these sensors, however, has been to awaken a generation of earth scientists to the possibilities of spaceborne microwave remote sensing, and many a student who carried out college exercises on interpretation of SEASAT and SIR-A data can now finally use these skills in real projects using the new generation of satellite sensors.

Figure 4.6. A Seasat image of part of the Isle of Wight and Solent, southern England. The effects of bathymetry on the surface roughness of the sea can be clearly seen (courtesy RAE).

4.1.11 ERS-1

This, the first operational civilian microwave satellite, was launched in July 1991, thirteen years after the failure of its direct predecessor, SEASAT. ERS-1 is designed primarily as an oceanographic satellite, and carries a suite of remote sensing instruments optimised for oceanographic observations. The improvements in weather forecasting, ship routing, and hazard prediction and monitoring for off-shore operations that ERS-1 is already bringing are of only indirect interest to most of the minerals industry, but this satellite will also provide information of more direct relevance. The Active Microwave Instrument (AMI) on board will permit SAR imaging of parts of the earth's surface, and this imagery, especially when used in conjunction with data from other sensors, will be of great value in mineral exploration, especially in structural studies. The lack of an on-board recorder, and power and programming constraints, mean that global coverage will not be available. There are many more ground stations in the ERS-1 programme than in the time of Seasat, and extensive areas of South America, Asia, and probably also parts of Africa, will be covered at some time during the life of the satellite by SAR imagery. Processing of raw SAR data to an image format has also improved dramatically since the days of SEASAT, and processing delays should not be a problem.

ERS-1 also carries other sensors which might occasionally have specific applications for the mineral industries. The radar altimeter enables very precise measurements of the water levels in large lakes and even the widest portions of some large rivers. This can provide warnings of flood or drought in remote parts of the world, and may even assist in assessments of water supply possibilities for major mining ventures. The scatterometer may enable regular measurements of soil moisture to be made at a very coarse resolution in remote parts of the world. This information will supplement the data already available from other meteorological satellites on rainfall in these areas, and thus assist in warning of floods and droughts, as well as in defining the probable mean annual rainfall in areas without direct observation. The Along-Track Scanning Radiometer (ATSR) allows very precise measurements of sea surface temperatures. These are not of direct interest to the mineral industry, but it is possible that the multispectral thermal capability will have similar regional geological applications as the AVHRR on the NOAA meteorological satellites. The 3.7 micron band on ATSR should also permit monitoring of forest fires and volcanic eruptions, as is possible with AVHRR. The characteristics of the various sensors on board ERS-1 are summarised in Table 4.12.

Table 4.12. ERS-1 sensors

1) AMI

Mode	Frequency	Polarisation	Incidence Angle	Spatial Resolution	Swath Width
Image Mode	5.3 ghz (C)	VV	23	30 m	100 km
Wave Mode	5.3 ghz (C)	VV	23	5 km	
Wind Scatt	5.3 ghz (C)	VV	variable	50 km	500 km

2)Radar Altimeter

Frequency	Spatial resolution	Altitude Accuracy
13.8ghz (Ku)	6.3 km	10cm

3) ATSR

Spectral Channels	Spatial Resolution	Swath Width
1.6, 3.7, 11, 12 micron	1km	500km

Since ERS-1 is primarily an experimental satellite, the orbit will be changed periodically during the mission to permit different types of observation. For this reason the observation repeat cycle will change. When the satellite is in a three day repeat mode, which will probably be the dominant mode during its life, global coverage with the scatterometer, altimeter and ATSR will be obtained every three days, and since the data rates of these instruments are relatively low, the data will be recorded on board for re-transmission to appropriate ground stations. This mode will not, however, permit complete coverage with the imaging radar. Sites along the track have the possibility of being imaged every three days, but sites between tracks will not be imaged at all during this cycle. A 35 day cycle permits complete SAR image coverage, and this should permit every site within range of a ground receiving station to be imaged at least once during the life of the satellite. At the end of its life, ERS-1 will go into a 176 day repeat cycle, which is designed to provide very closely spaced radar altimeter information in order to refine the definition of the mean sea surface and ocean geode, an activity which is not likely to cause great excitement in the mineral industry, although if the imaging SAR operates during this final stage, the considerable overlap between images may permit some stereoscopic studies, although the base/height ratio would be small.

4.1.12 Airborne Scanners

Airborne scanners are used for three main purposes. One is largely experimental, to use a sensor on an aircraft platform to obtain information about spectral responses of surfaces and materials, and to test scanning systems which may later be improved, or in some cases simplified, for use in spacecraft. The Daedalus 11-channel multispectral scanner was widely used to obtain test imagery of the type to be acquired later by the TM on Landsat 4. This imagery was used to evaluate the possible applications of TM, and to develop image processing techniques appropriate to this multi-band imagery. The TIMS multispectral thermal scanner is intended to evaluate the applications of multi-band thermal imagery, especially for geological mapping and mineral exploration purposes. Positive results might lead to the incorporation of a multi-band thermal capability on future earth observation satellites. Various imaging spectrometers (AIS, AVIRIS, GER) have been flown in a series of international campaigns to investigate the uses of multi-band narrow spectral resolution remote sensing. Some campaigns have been organised to simulate the imagery to be obtained by specific future spaceborne systems, for example on the Japanese JERS-1 and ADEOS satellites, while others have been directed towards identifying the specific wavebands amongst the 228 (on AVIRIS) which have the most information content and discrimination power for specific applications. This information will be used to plan future spaceborne instrumentation, and to schedule the operation of spaceborne spectrometers.

A second major use of airborne systems has been to acquire imagery at specific times on demand. Operational constraints with spaceborne systems result in compromise overpass times that do not suit all applications, and a spaceborne system cannot readily respond to emergency situations. This is especially true of thermal imaging. The specific heat and thermal emissivity of rocks can be an important diagnostic feature, but differences are often obscured in day-time by topographically influenced solar heating. Imagery acquired just before dawn can be especially valuable, but no fine-resolution spaceborne system acquires imagery at that time. Airborne thermal scanners can be flown at whatever time best suits the application. Such features as oil slicks and discharge of other pollutants from ships or factories may be relatively short-lived and localised phenomena, only detectable at specific times that do not coincide with satellite overpasses. Airborne scanning can be scheduled at relatively short notice to monitor these pollution events. In cloudy climates, the acquisition of spaceborne imagery within specific time windows, critical for monitoring crop growth, for example, is often very much a matter of chance. Airborne scanners can be used to acquire imagery on cloud-free days within these time windows, and can be more flexible in their flight times to take advantage, for example, of sunny afternoons following cloudy mornings.

Table 4.13. Characteristics of some typical airborne scanners

Scanner	IFOV	Scan angle	Wavelengths	Notes
Daedalus	2.5 mrad	86	0.42 - 0.45	visible
AADS 1268			0.45 - 0.52	
			0.52 - 0.60	
			0.61 - 0.63	
			0.63 - 0.69	
			0.70 - 0.75	visible
			0.76 - 0.90	near infrared
			0.91 - 1.05	near infrared
			1.55 - 1.75	mid-infrared
			2.08 - 2.35	mid-infrared
			8.50 - 13.0	Thermal
Moniteq	variable	70		
Spectral mode			128 channels	Only in
40 pixel swath			between 0.40	visible
			and 0.80	and near
				infrared (CCD's)
Spatial Mode			Eight selected	
1925 pixel swath			channels between	
			0.40 and 0.80	
TIMS	2.5 mrad	80	8.2 - 8.6	
			8.6 - 9.0	Six
			9.0 - 9.4	channels
			9.4 - 10.2	in thermal
			10.2 - 11.2	region
			11.2 - 12.2	

Plate 1. Landsat MSS false-colour composite of an area in northern Kenya. Volcanic landforms and dark unvegetated lavas and pyroclastics can be seen in the Rift Valley (left), while highland areas are densely vegetated (right). Geological structure is most clearly seen in the sparsely vegetated areas. (Courtesy EOSAT)

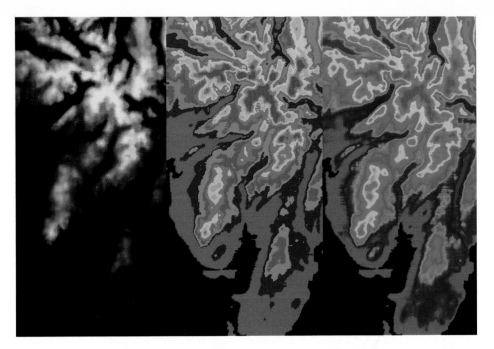

Plate 2. An illustration of density slicing. A digital elevation model (left) has been density sliced using different techniques. A simple slice (centre) assigns solid colours to each height interval, while a more complex slice (right) preserves more detail within the height intervals.

Plate 3. Part of a Landsat TM scene of central Kenya. This false-colour composite assigns red to band 4 (NIR), green to band 5 (MIR) and blue to band 3 (visible). This combination highlights vegetation differences. Natural forest appears reddish brown, grassland is green and bare soil white. Irrigated crops (top left) are shades of red. Note the major WNW fractures associated with the Rift Valley, and volcanic landforms. The active Menegai volcano is at upper left. (Courtesy EOSAT)

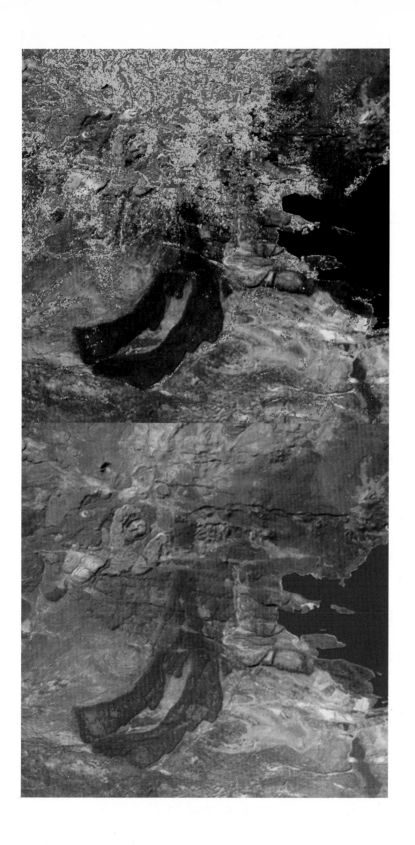

Plate 4. An example of vegetation indices. A Landsat TM extract of an area in the north-central Rift Valley of Kenya appears at left. Numerous volcanic features (cinder cones and lava flows) can be seen north of Lake Baringo. Vegetation appears in reddish tones. The right image is a colour-coded vegetation index superimposed on a monochrome image. The NDVI is colour coded from densest vegetation in red to sparse vegetation in green. (TM imagery courtesy EOSAT)

Plate 5. The use of decorrelation stretching to improve lithological contrast. The TM extract on the left, showing volcanics in the north-central Rift Valley of Kenya, has been decorrelation stretched using intensity - hue - saturation transforms to produce the image on the right. Distinction between lithologies is greatly improved. (TM imagery courtesy EOSAT)

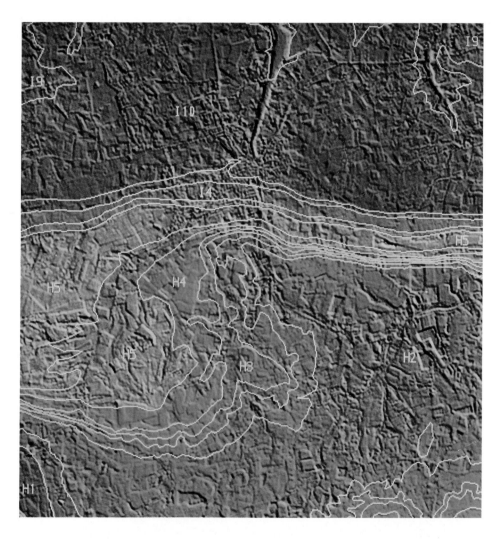

Plate 6. The use of intensity - hue - saturation transforms in data integration. A filtered satellite image (intensity) has been combined with a digitised geological map (hue) and a digital elevation model (saturation) to produce this composite image which enables the interpreter to relate structural features to lithology and elevation in hydrogeological studies.

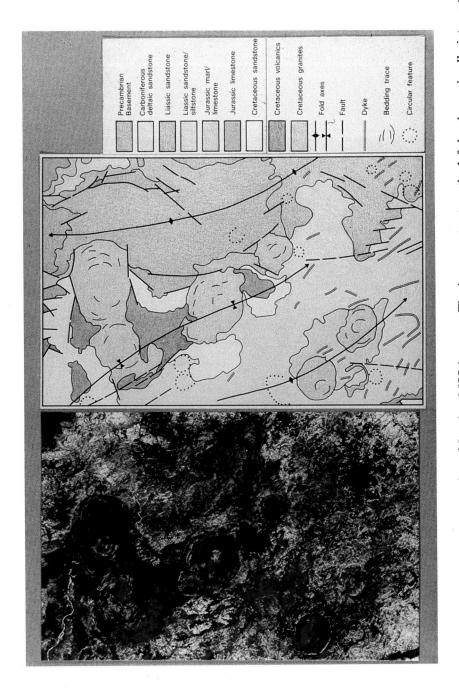

Plate 7. An example of geological interpretation of Landsat MSS imagery. The image extract on the left has been visually interpreted, with reference to previous geological sketch maps, to produce the new map on the right. (Courtesy Hunting Technical Services)

Precambrian Basement

Carboniferous deltaic sandstone

Liassic sandstone

Liassic sandstone/ siltstone

Jurassic marl/ limestone

Jurassic limestone

Cretaceous sandstone

Cretaceous volcanics

Cretaceous granites

Fold axes

Fault

Dyke

Bedding trace

Circular feature

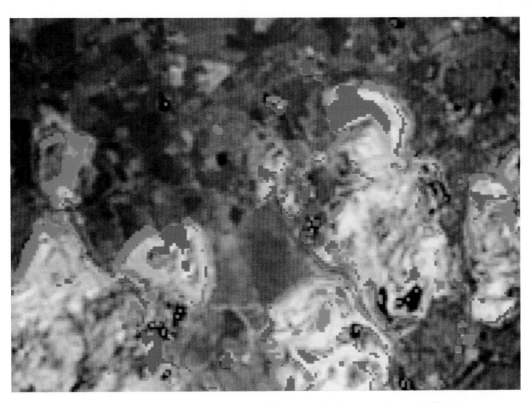

Plate 8. Change detection in a mining district. Digital processing of TM imagery acquired in 1985 and 1988 resulted in production of this image showing change in the extent of mine workings and dumps. Red areas indicate increase, mainly in waste dumps, while green indicates decrease, mainly as a result of vegetation cover on flanks of dumps. St. Austell district, Cornwall.

Plate 9. The use of multispectral thermal imagery in geological mapping. A coarse resolution AVHRR image of an area in southern Morocco has been processed to enhance lithologies. The image on the left combines a band 2 image (intensity, for background detail) with a 5/4 thermal ratio image (hue, for lithology) and a 2/1 ratio (saturation, for vegetation cover). The close correspondence with the geological map on the right is evident.

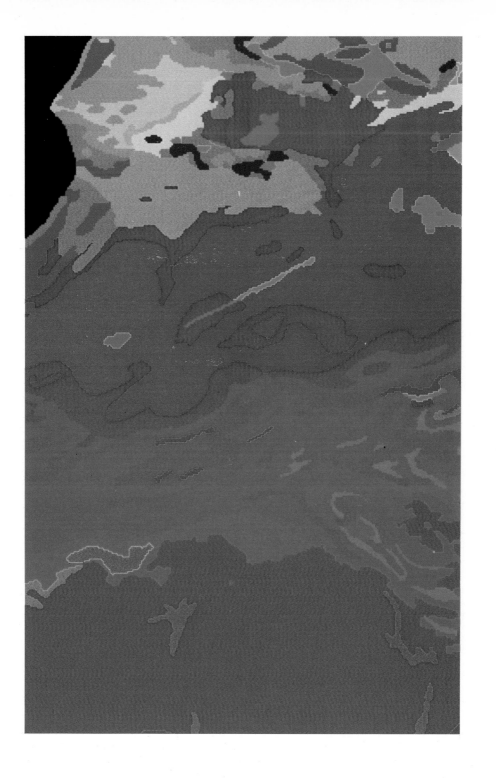

Plate 10 A directed principal components colour composite image of the Roberts Mountain area, Nevada. Areas of intense "hydroxyl" alteration appear red, and intense iron oxide alteration bright blue. Where both alteration types coincide the image is white. Note the coincidence of the highly altered zones with locations of mines and prospects shown in Figure 5.1. (Courtesy Loughlin)

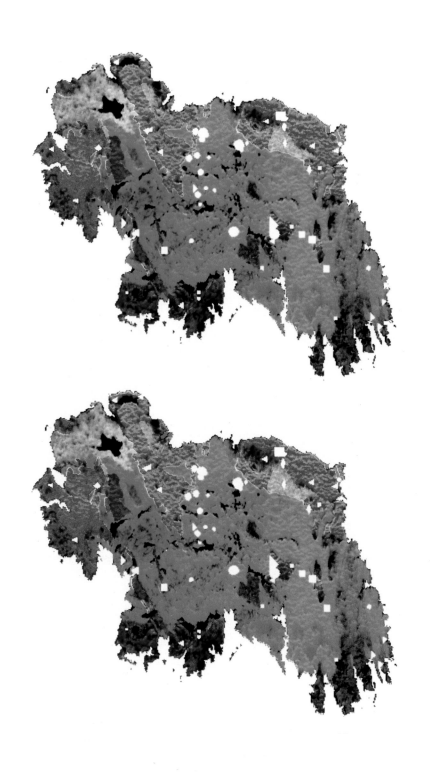

Plate 11. A digitised geological map of Ireland (hue) has been combined with an AVHRR band 2 image (intensity) and a filtered AVHRR thermal image (saturation). This colour composite was then transformed into a stereo pair, using a digital gravity map as elevation. Mineral deposits are indicated by white symbols, and stereo viewing enables of the lithological and structural setting of these deposits.

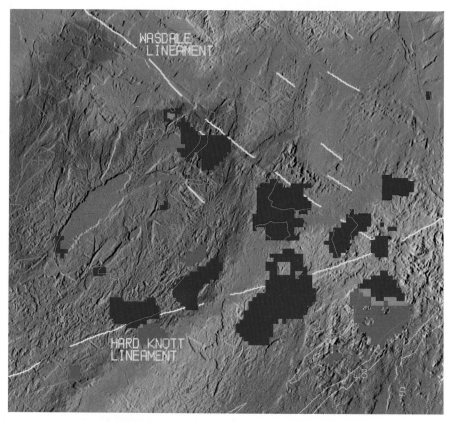

Plate 12. Data integration for mineral exploration. A filtered TM image (intensity, for structure) has been combined with a gravity image (hue). Saturation is set to a fixed value. Geological boundaries are superimposed in pale blue, interpreted major lineaments in white, and geochemical anomalies for tin and lead/zinc in red and green respectively. English Lake District. (Courtesy Harding and Forrest)

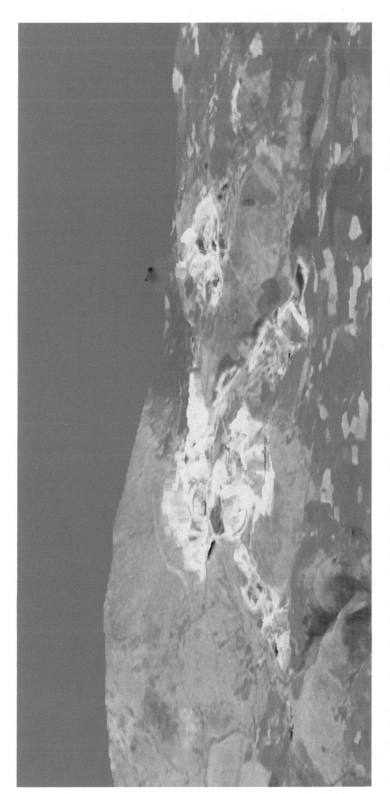

Plate 13. A three-dimensional view of the Lee Moor china clay workings in south Devon, produced from 1984 Landsat TM imagery combined with digital elevation model.

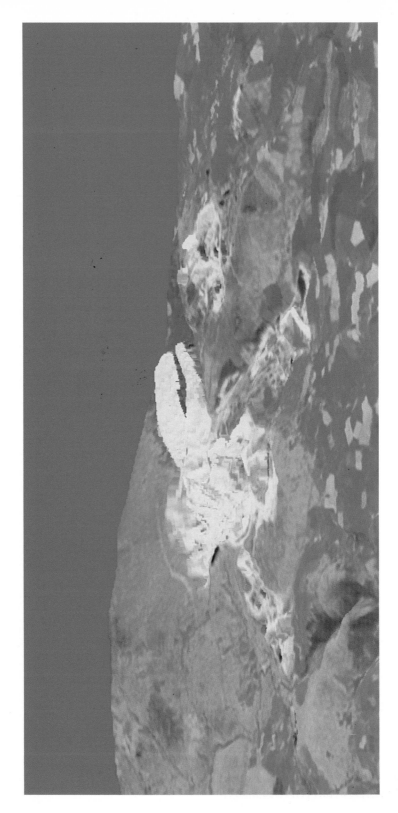

Plate 14. The same view as Plate 13, but with changes to pits and dumps added to imagery and digital elevation model to simulate the visual effect of a possible mining plan.

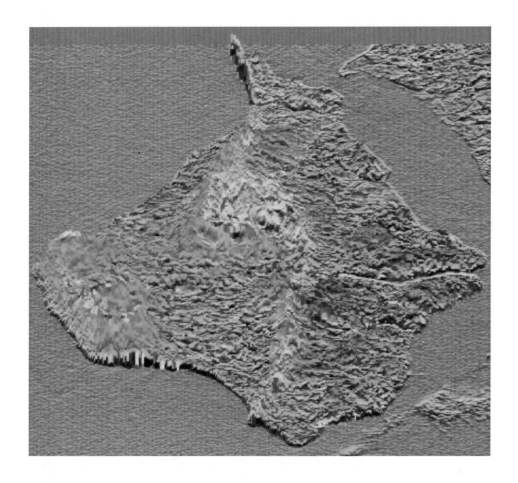

Plate 15. A perspective view of the Isle of Wight, combining a geological map, filtered satellite imagery and a digital elevation model. The monoclinal geological structure is clearly visible in this enhanced image.

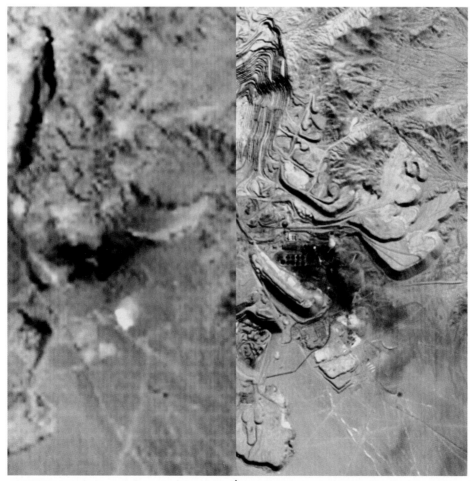

Plate 16. Two views of part of the Chuquicumata copper mine in Chile. The image on the left is a 1973 Landsat MSS image, while that on the right was acquired in 1987 by SPOT. Changes in dumps can be seen despite the poor spatial resolution of the MSS image. The improvement in detail visible with SPOT is also apparent, especially for features such as the concentrator. (Copyright CNES 1987)

The third main reason for using airborne scanners rather than spaceborne systems is the improvement in spatial resolution which is possible from an aircraft. The spatial resolution of airborne systems is not fixed, but depends on the flying altitude above the ground surface. Each scanner has a fixed instantaneous field of view (IFOV) and scan angle. The IFOV is the angle subtended at the scanner by one image pixel on the ground. Since the angle is fixed, the pixel size will increase as the aircraft height increases. The scan angle is the angle from one end of a scan line to the other. The width of area imaged will thus increase with aircraft height. In any airborne scanner imaging, there is a trade-off between spatial resolution and area covered. The finer the spatial resolution, the narrower the swath imaged on each flight line, and thus the greater the cost per square kilometre. If very detailed imagery is required, for example of a series of mineral prospects rather than a whole exploration area, this is usually flown over a series of small areas, since it would be uneconomic and probably unnecessary to fly very large areas at fine resolution.

Airborne scanners are basically similar in design to those in spacecraft. There are optomechanical (for example, Daedalus) and pushbroom (for example, Moniteq) varieties. The actual detectors, especially those operating in the mid-infrared and thermal portions of the spectrum, must be cooled to very low temperatures, and data is usually recorded digitally on magnetic tape. The scanners range from very simple single-channel thermal infrared systems (Infrared Linescan) to hundreds of channels in some experimental airborne spectrometers (AVIRIS). Some systems have only a relatively small number of channels, but the wavelengths imaged are programmable in advance so as to acquire imagery relevant to the specific application (Moniteq). For example, studies of marine environments might require numerous wavebands of imagery in the visible portion of the spectrum, while vegetation studies require imagery in the near-infrared and visible. One scanner (the Australian Geoscan) was specifically designed for mineral exploration purposes, with a multispectral capability in the important 2.2 micron region of the mid-infrared. Aircraft are unfortunately not such stable platforms as spacecraft, especially when operating at low altitudes. In most airborne scanners the actual scanner unit is mounted in gyro-compensated gymbals so as to minimise the effect of aircraft roll, yaw and pitch, but it is rarely possible to eliminate all effects of aircraft motion in low-altitude airborne scanning. The wide scan angle, commonly more than 80 degrees, compared to less than 20 degrees in most spaceborne scanners, results in airborne scanner imagery being much more distorted than spaceborne imagery. Pixels towards the edges of scan lines cover larger distances on the ground in an across-track direction than those near the centre of the swath, but this can be compensated for during later processing. Topography induces distortion which is more difficult to correct, especially in low-altitude imagery.

Comparative studies of Daedalus airborne scanner imagery with TM imagery of test sites in the western United States and Saudi Arabia have shown that, for mineral exploration purposes, TM could locate the same targets at much lower cost. If imagery is required at wavebands currently not obtainable from space, or if there is a requirement for very detailed imagery of mineral prospects, then airborne scanners will have to be used. In many respects, airborne scanners can be regarded as a transitional phase in remote sensing. They permit a preview of new types of imagery that will eventually be available on a global basis from space. Even when spaceborne scanners of equivalent spectral and spatial resolution are available, there will still be a place for airborne remote sensing as a source of imagery of small areas at times chosen by the user. The cost of airborne scanning depends very much on the size of area to be flown, the remoteness of the region, and the local weather conditions. As a rule of thumb, multispectral airborne scanner surveys cost in the range of ten to one hundred times as much per square kilometre as satellite imagery. The characteristics of some typical airborne scanners are presented in Table 4.13.

4.2 FUTURE ENHANCEMENTS AND NEW DEVELOPMENTS

Satellite remote sensing is a very rapidly growing science, and has come a remarkably long way in the relatively few years since the launch of the first Landsat satellite. Books on remote sensing have a rather short shelf life, since portions of them at least are rapidly rendered out of date by developments in sensor, processing and applications technologies. In attempting to summarise planned improvements and diversification in sensors, and also to speculate on longer-term future developments, I am very conscious that, if people in the minerals industries purchase this book, many of the things that I say in this chapter may either have already happened or have been superseded by other new discoveries by the time that some people read this.

The main trends in sensor development over the past ten years or so have been in four main directions. Firstly there has been an increase in the number of wavebands of data available, and in the number of portions of the electromagnetic spectrum imaged. The original four wavebands of the Landsat MSS have been expanded to seven in the TM, with new information from the mid-infrared and thermal portions of the spectrum. Airborne imaging spectrometers now collect data in hundreds of channels simultaneously. A multispectral thermal capability has become available from the AVHRR and ATSR sensors, as well as from airborne scanners. New radar systems have covered different portions of the microwave part of the spectrum. The second major trend has been towards increasing spatial resolution. The seventy metre pixels of MSS have been succeeded by the thirty metres of TM and the ten metre pixels of SPOT PAN. A third direction of improvement has been the addition of steerable sensors, notably in SPOT, which allow the generation of stereoscopic imagery and the more frequent acquisition of imagery. Finally there has been a steady increase in the number of operational remote sensing satellites, from one non-meteorological system in the 1970's to twelve in 1991.

It is likely that these trends will continue for the next decade, although not every application requires improved spectral or spatial resolution, and some can be more efficiently performed with simpler imagery. Later SPOT satellites will have at least one extra waveband, imaging in the critically important 1.5 micron portion of the spectrum. Future Landsat satellites may also have additional wavebands, and will certainly feature finer spatial resolution and a stereoscopic capability. Special purpose sensors such as those on JERS-1 and ADEOS will have enhanced multispectral capabilities in geologically important portions of the spectrum. A series of experimental instruments planned for launch on polar platforms in the second half of the 1990's may include spaceborne spectrometers, observing portions of the earth in hundreds of spectral channels, and multi-wavelength, multi-polarisation imaging radars, bringing the possibility of all weather observation with similar discrimination powers to today's optical systems. The increase in the total number of satellites will continue, hopefully in coordinated orbits, so as to increase the possibility of image acquisition at critical times such as those of natural disaster, and to allow the collection of time series for monitoring purposes. Enhanced meteorological satellites, with finer spatial resolution and observations at more wavelengths than today's geostationary and polar-orbiting platforms, will offer greater possibilities for use by non-meteorological disciplines, particularly in monitoring roles observing fires, floods and famines.

A large amount of the data collected by today's earth observation satellites is either redundant or never consulted. Many systems image continuously, and all transmit the entire essentially unprocessed image back to earth, where it is recorded and archived, possibly never to be examined again. The SPOT satellites started the trend towards programmed image acquisition, and a significant proportion of each SPOT orbit is now occupied with acquisition of imagery on demand, rather than speculatively for archive. This trend is likely to strengthen in the future, particularly with the urge towards self-sufficiency that the remote sensing industry is facing from previously generous government funders. The financial advantages in only acquiring imagery on demand are obvious, but so are the dangers. Many breakthroughs in remote sensing applications development have been made using imagery acquired for other purposes, and in many environmental and even legal problems it may be essential to have access to old imagery in order to monitor change. The question of data redundancy is also being addressed. On-board processing of data on satellites is now possible, thanks to the increased speed and decreased size and power requirements of new generations of computers. Data compression techniques have been developed to dramatically reduce the volume of data to be transmitted down from the spacecraft and later stored by receiving stations. "Smart" sensors could look forwards along the spacecraft track, and steer the main sensors so as to image areas free from cloud. It has even been suggested that change detection studies could be carried out on-board, with a newly acquired image being compared to a stored one in real time. An image showing no change would not be stored or transmitted to ground. Programmable spectrometers could be used to acquire only those few critical wavebands needed for a particular application, instead of acquiring hundreds of non-informative bands of data. Future large platforms may allow simultaneous observation with a range of different sensors, allowing better co-registration of, for example, microwave and optical imagery because of the lack of temporal difference. On-board processing could also result in the generation of derived products occupying a much smaller data volume than the raw data from which they were generated. A very simple example could be the generation of a vegetation index image by combining infrared and red images, but the derived products could be much more complex, and reduce data volumes even more.

One possible development that could be of special interest to the minerals industry and others working in remote areas is the idea of acquiring relatively small images, in terms of their area of coverage, to customer order, and then transmitting them down directly to the user in a form permitting reception on relatively small and low cost antennas. A geophysical crew working in a remote area, for example, might encounter unexpected access problems to a prospect. They could put in a programming request through a communications satellite link, and receive a detailed and up to date image of their site within days. This process could also ensure confidentiality, since the image need not be recorded on the spacecraft or transmitted down to any groundstation other than the portable one of the customer.

4.3 CHOOSING THE APPROPRIATE SYSTEM
The newcomer to remote sensing can be forgiven a certain bewilderment when faced with the present range of remotely sensed imagery which is available. His situation is not helped by the vested interests of many consultants, who can often also be distributors for one or other of the satellite operators. The choice of an appropriate sensor for a specific task depends on a variety of factors, including cost, spatial resolution, spectral resolution and frequency of acquisition. Studies of a very large area may be most effectively carried out with different imagery than a very detailed study of a single mine site. In general, the cost per unit area of satellite imagery varies inversely with the spatial resolution. The finest spatial resolution digital imagery currently available is that from the SPOT panchromatic sensor, and this costs approximately fifty times as much per square kilometre as one band of Landsat TM imagery, more than one hundred times as much as Landsat MSS, and about fifty thousand times as much as imagery from the AVHRR sensor on the NOAA satellites. It would obviously not make economic sense to use SPOT PAN imagery to map geological structure on a continental scale, when the features of interest are tens, hundreds or thousands of kilometres in length, but equally the spatial resolution of AVHRR imagery is so coarse than few mines would even fill a single pixel of the imagery. Mapping land use around a proposed new mine could not realistically be done with panchromatic imagery, since spectral information is essential to the discrimination of vegetation types. Sensors which obtain information from the infrared portion of the spectrum are essential for vegetation mapping, and for many studies it is preferable to have information from the mid-infrared as well as from the near infrared. At present, this dictates the use of Landsat TM, despite its poorer spatial resolution than SPOT. Mapping of lithologies and alteration zones likewise requires information from the mid-infrared, and thus must use TM, although structural studies do not usually require multispectral imagery, and can be carried out with whichever imagery is most appropriate to the scale of the investigation. Most applications in the minerals industry do not require frequent acquisition of imagery, and for many applications, particularly in mineral exploration, imagery acquired many years ago is almost as useful as that acquired yesterday. Monitoring of disasters, or of regional seasonal changes in vegetation, does however require frequent, or at least timely, imagery, so that the choice of imagery may be decided by the acquisition frequency.

Table 4.14 is intended merely as a guide for the inexperienced. It is based mainly on the author's personal experience and prejudices, and may be disagreed with vehemently by others with greater experience in particular fields. It should however help to save new users of remote sensing from the worst mistakes, and following these suggestions is unlikely to result in purchase of totally unsuitable data.

Table 4.13. Recommended sensors for different tasks

Application	Scale	Recommended Satellites and Sensors
Structural Geology	continental >1:2 million	Geostationary meteorological satellites
	Regional >1:1 million	NOAA AVHRR, Band 2 (NIR) imagery
	Regional 1:250,000	Landsat MSS, Band 4 (7 for Landsats 1,2,3)
	Local 1:50,000	Landsat TM band 4 (sometimes 5 or 7)
	Local 1:20,000	SPOT XS or SPOT Pan, former band 3
Lithological Mapping	Regional >1:1 million	NOAA AVHRR thermal and NIR (bands 2, 4, 5)
	Local 1:50,000	Landsat TM (all bands, sometimes including thermal)
	Local <1:20,000	Airborne scanners (Daedalus or others)
Logistics	Regional 1:200,000 to 1:1 million	Landsat MSS (all bands)
	Local 1:25,000 to 1:100,000	Landsat TM (bands 3,4,5 or 4,5,7)
	Local <1:50,000	Spot PAN or PAN/TM composite
Land-Use Mapping For EIA etc	Regional and Local 1:50,000 to 1:200,000	Landsat TM (all bands except thermal)

5

Remote Sensing and GIS in Mineral Exploration

Remote sensing is not the answer to every explorationist's prayer. It is not a magic "black-box" that will produce a map with a neat arrow labelled "dig here". Like the other relatively new technologies of geophysics and geochemistry, it can be of great assistance to the explorationist in the increasingly difficult task of finding new orebodies, but it can only rarely do this as a stand-alone technique. In the early days of geophysics, in particular of airborne electromagnetic systems, many new orebodies were discovered with little assistance from other techniques. Areas of promising geology, often in previously known mining districts, were covered with deep glacial deposits which rendered conventional exploration techniques ineffective. Airborne EM was able to locate ore deposits under this cover, and great discoveries were made. Soon the most obvious deposits had all been found, and increasingly sensitive geophysical systems are now used in conjunction with a whole range of other techniques to explore for deeper and less conductive deposits. The same is true for geochemistry. In certain favourable environments, geochemistry alone has located important mineral deposits, but in most well-explored districts, where any remaining deposits will by definition be well hidden, geochemistry is just one tool amongst many. In some parts of the world, in particular arid areas with significant relief, remote sensing can, on its own, identify alteration zones around certain types of mineral deposit, and has been credited with the discovery of some mines. Most deposits are however more subtle, and few environments in the real world are as simple as those used in tests of new systems. Remote sensing must therefore be used intelligently, as appropriate to both the desired target and the geological, climatic and topographic environment, as yet another powerful tool in the exploration geologist's kit. The main fields of application are in logistics, mapping structure and lithology, and in the location of alteration zones. Airborne scanner imagery can also be used in the detailed mapping of mineral prospects. At all levels of a mineral exploration programme, remote sensing can be of greatest benefit when used in conjunction with other more conventional sources of information, ranging from topographical and geological maps through production information to geophysical and geochemical data.

5.1 LOGISTICS

The use of remote sensing as a logistical aid in mineral exploration has become so widespread that many users do not even realise that they are employing satellite imagery. False-colour composite imagery has been in regular use as a crude but up to date map of remote areas since the mid-1970's, and although more sophisticated products, derived from Landsat TM or Spot rather than MSS, geometrically corrected to standard map projections, and sometimes even annotated with placenames and geographical grids, are now in fairly common use, the simple uncorrected photographic image product is still almost certainly the most familiar remote sensing aid to most field geologists and geophysicists.

The reasons for the early acceptance and widespread use of satellite imagery as a logistical aid in exploration are easy to find. Much mineral exploration is carried out in remote areas of the world which are often poorly mapped, and there is thus a need for base maps in order to plan any exploration programme. Even where good maps exist, the pace of change in these remote areas over the past twenty years is such that they rapidly become out of date. Even though rivers tend to keep to their courses, and mountain ranges remain in place, patterns of settlement and communications change very rapidly. What might have appeared as untouched rainforest on maps produced in the 1960's may now be a landscape of peasant farms, and the logistical problems of transporting men and equipment into rainforest, and of working in the forest, are replaced by the problems of having to negotiate access rights with hundreds of individual landowners. Desert areas without permanent human settlement or sources of surface water can be transformed to irrigated farmland, removing problems of water supply and access for field work, but greatly increasing the difficulty of actually carrying out the work. Government agencies may often be poorly informed about these changes, or even be unwilling to discuss them with foreigners, and the need for accurate and timely information on access and land cover types can often only be met by the remote sensing satellite, whose observations are not restricted by local or national boundaries, and whose imagery can provide quantitative map-like information.

The user should not be misled, however, into thinking that satellite images are the same thing as maps. Topographical maps are the result of interpretation and synthesis of a range of different observations, both on the ground and, in many cases, by means of air photography. The satellite image, even a processed and geometrically corrected false-colour composite, consists of the raw data or observations, without interpretation or synthesis. The image can often show features which do not appear on a map, either because these are not easily visible to the surface observer, forced as he is to see things without the benefit of infrared wavelengths, or because the features occur in places not easily accessible to the ground observer, especially before the days of routine use of helicopters in mapping projects. In some cases the map may lack features prominent on satellite imagery because of deliberate censorship by the map-maker. The satellite image is essentially a source of raw information about the land surface. Interpretation and synthesis can be carried out by the individual user, assisted perhaps by superimposed placenames and geographical grid, or the image can be used as a base for a new interpretative map. The approach adopted will depend both on cost and on the use to which the information is to be put.

A preliminary reconnaissance of an unknown area is unlikely to justify more than a standard false-colour composite, purchased from a distributor. This can be used as a source of basic information about the area, and serve as a planning base for the first stage of field work. More intensive field work will usually justify the production of multiple copies of specially processed imagery. The use of different band combinations and special filters can enhance those features which are of particular interest to the project, and geometric correction is often desirable at this stage. If the organisation is large enough, and the duration of expected use of the imagery sufficiently long, it may even be economic to produce colour printed image maps, where the unit cost is much lower than photographic products provided that numbers in excess of at least 500 are required.

The Directorate General for Mineral Resources in Saudi Arabia makes very extensive use of satellite-derived map-like products in its mapping and exploration programmes. Initially, single-band (monochrome) optically processed mosaics of Landsat MSS imagery were produced at scales of 1:250,000 and 1:100,000, the former for aircraft navigation in the desert and the latter as mapping bases for recording information from geological reconnaissance. This was necessitated initially by the lack of any topographical maps for much of the kingdom, and even when accurate 1:50,000 topographic maps were produced, security considerations did not permit their routine use by foreign geologists and pilots. The early Landsat products were followed by a series of 1:250,000 scale colour image maps, based again on MSS imagery, but digitally processed and geometrically corrected, with superimposed place names and geographic grids. These have been in routine use, particularly for helicopter navigation, since 1984, and most pilots much prefer them to conventional maps, since the image maps bear a much closer relationship to what the pilot can actually see of the land below him than does any interpretative map.

The type of imagery required for logistical purposes depends to some extent on the size of the area to be covered, but it is usually desirable to have the finest resolution affordable. Since the imagery will often be used to locate access tracks, bridges, and similar man-made features, the use of resolution which will unambiguously display these features is important. Colour imagery is also almost essential, since it allows discrimination of densely vegetated land and areas of water. For very large and remote areas with little or no human settlement, Landsat MSS imagery may still be adequate, but in all other cases the higher cost of TM or SPOT imagery would probably be justified by the increased information content. SPOT panchromatic imagery on its own is not recommended for logistical use, because of the poor surface discrimination, and for many projects the spatial and spectral resolution of Landsat TM imagery would appear to be ideal. If the exploration areas are smaller, or if the area is densely settled and farmed, the threefold increase in cost per square kilometre of SPOT multispectral imagery would be justified, while in extreme cases where the finest possible spatial resolution and best spectral contrast is required, and cost per unit area is not a pressing consideration, combinations of SPOT panchromatic 10 metre resolution imagery with

Landsat TM could be used. Some of the most impressive image maps in the world are produced by combining these two sensors using proprietary resampling techniques developed by the largest remote sensing processing companies. Very sharp image maps in a full range of natural appearing colours can be produced at scales as large as 1:20,000, and are beginning to compete with conventional maps in many fields far removed from mineral exploration. Since any mineral exploration programme is in essence a process of progressive elimination of less promising ground and the retention of only the most promising, and often widely scattered portions, the cost of regional coverage with such attractive products is rarely justified, although they could be of great value during detailed exploration and prospect evaluation stages of a programme.

5.2 REGIONAL GEOLOGICAL MAPPING - LITHOLOGY AND STRUCTURE

Most geological maps produced during the past thirty years have included some photo-interpretation, and many reconnaissance maps of remote areas of the world have been produced almost entirely from air photos. The process of preparing a geological map consists in most cases of scattered ground observations of rock outcrops, followed by what is often a highly subjective interpolation between the observation points. The interpolation procedure is normally guided by air photo-interpretation. In many parts of the world the collection of surface data is restricted by lack of natural rock outcrops, and it may even be necessary to create artificial outcrops by pitting, trenching or drilling in cases where there are severe ambiguities in the geological interpretation, or where critical exposures are lacking. Even in areas of considerable exposure, time rarely permits the detailed mapping of all rock outcrops, except in detailed studies of small areas at scales of greater than 1:10,000, and interpolation is thus a major part of any geological mapping.

The traditional use of stereographic monochrome air photos has been for structural information, with some lithological guides. Faults and other linear features are interpreted, as are prominent curvilinear features which may follow the outcrops of resistant beds. In the simplest case, two observed outcrops of resistant quartzite may be seen on the air photos to be connected by a prominent ridge, and can thus be mapped as a continuous horizon of quartzite. Mapping of lithologies from air photos depends mainly on variations in vegetation cover, although surface roughness variations can also be important if they occur on a large enough scale. Unique identification of rock-types or mineralogies, without field information, is rarely possible from air photos. The photos are commonly used at least twice during a mapping exercise, firstly to define the overall structure and identify key locations for ground checking, and then to aid interpolation after the main phase of field work. A further iteration of field checking and photo interpretation is often necessary after the first interpolation, in order to check problem areas identified. All this work is based on visual interpretation by experienced geologists, preferably those actually carrying out the field work. It is labour intensive, but many geologists find that time spent on photo-interpretation is extremely valuable, since it aids in refinement of the geology of the area, and stimulates thought about alternative interpretations.

Satellite remote sensing has not replaced conventional air photo-interpretation in most geological mapping programmes, although it is playing an increasing role in many areas. The scarcity of stereo satellite imagery and the relatively coarse resolution of much current imagery are two important factors preventing total replacement of conventional air photography by remote sensing. There is also an important question of cost. Although it is vastly more expensive to acquire new air photography over a large area than to purchase satellite imagery, much geological photo-interpretation does not require the most up to date photography. Most national geological surveys and major mining companies already have extensive archives of air photos covering their main areas of interest, and photos not already held in archive can usually be purchased at fairly low cost from national survey agencies. Geologists are also accustomed to taking air photos, plus a pocket stereoscope, with them into the field for interpretation in camp, or even while sitting on an outcrop. Satellite imagery has normally been regarded as too expensive for this sort of use, and is usually only available to the field geologist at well-established field camps or back at headquarters. There is also the problem that the only satellite imagery that many geologists have seen, even ten years after the launch of Landsat 4 and five years after SPOT, is Landsat MSS imagery from Landsats 1, 2 and 3. The coarse spatial resolution and lack of lithological discrimination of this imagery gave a falsely negative impression of the capabilities of remote sensing, and hardened many field geologists in their prejudice that old-fashioned air photo-interpretation was the best way to go.

Modern satellite imagery can contribute greatly to geological mapping. It can do all that air photography can, except at the most detailed scales (greater than 1:20,000). It can be used exactly as conventional air photography, with monochrome stereo prints for visual interpretation with conventional stereoscopes, but it can also provide much more because of its multispectral nature and the powers of digital image processing.

5.3 ALTERATION ZONES

Many mineral deposits have extensive alteration zones associated with them, occupying volumes which are sometimes two or three orders of magnitude greater than the actual mineral deposit. Since such alteration zones present much larger targets than their associated mineralisation, they are often used as guides to mineralisation. There are many types of alteration zones. Some are geochemical only, without any change in the bulk chemistry or mineralogy of the affected rocks. These may be excellent targets for stream sediment and soil geochemical surveys, but are unlikely to be detectable by remote sensing unless some of the trace elements enriched in the alteration zone are highly toxic to plants. Other alteration zones are characterised by volumetrically small mineralogical modifications, for example the conversion of goethite to magnetite, or pyrite to pyrrhotite, which have dramatic effects on the geophysical response of the zone, but are usually not detectable by remote sensing. A third type of alteration zone shows major mineralogical changes, for example sericitisation of feldspars and introduction of iron in the form of oxides and sulphides. These mineralogical modifications are often exaggerated in the zone of weathering, since the altered rocks are often more susceptible to chemical weathering than their unaltered counterparts. These are the alteration zones which can, in arid and semi-arid environments, be located using satellite remote sensing.

Alteration zones of this type are associated with many classes of mineral deposits, but are particularly well-developed around hydrothermal deposits and volcanic-associated sulphides. Similar zones sometimes surround sedimentary sulphide deposits, although these are not strictly speaking the result of alteration, but rather of chemical sedimentation in an anomalous environment. The use of these alteration zones is especially important in exploration for hydrothermal gold mineralisation, since a valuable orebody can be an extremely small target, but may be surrounded by relatively large volumes of alteration. This class of mineral deposit is often a poor geophysical target, mainly because of the small size, and although there may be a distinctive geochemical signature, stream sediment geochemistry is not always effective in an arid environment.

The main characteristics of these alteration zones which are susceptible to detection using satellite remote sensing are a general increase in the overall reflectance, or albedo, the presence of ferruginous staining, and the presence of characteristic assemblages of clay minerals. The first two of these have been detectable, when sufficiently intense or covering a sufficiently large area, since the days of the early Landsat satellites. The multispectral scanner, with its four wavebands in the visible and near infrared portion of the spectrum and its spatial resolution of 80 metres, was able to detect large areas of bleached rock or weathered material, which had a much higher reflectance at visible wavelengths than the surrounding rocks. Band 5 of the MSS (2 on the same instrument when flown on Landsats 4 and 5) is in the red portion of the spectrum, and thus sensitive to ferruginous staining, which normally turns rocks and their weathered products red. A whole range of ratios were devised by mineral explorationists during the 1970's in an attempt to highlight these ferruginous zones in MSS imagery. Some of these performed quite well in arid environments, although they were not able to distinguish iron staining due to mineralisation from the much more pervasive staining resulting from the processes of lateritisation which are so common in many arid and semi-arid terrains.

The launch of Landsat 4 in 1983, carrying the TM scanner with its greatly increased waveband coverage and finer spatial resolution, and in particular with the 2.2 micron band (band 7), which was added mainly at the urging of the geological community, greatly increased the chances of detecting alteration zones using satellite imagery. Many exploration companies also used the Daedelus airborne scanning system, popularly known as the airborne thematic mapper (ATM), to experiment with remote sensing over known areas of mineralisation, and ultimately to carry out exploration campaigns over new areas. A wide range of techniques have been developed for recognition of alteration zones using the TM and ATM. As an example of one possible approach, I shall describe the technique developed by W. P. Loughlin at the United Kingdom National Remote Sensing Centre. I am grateful to him for allowing me to use his work in my book, and for providing the excellent illustrations of his technique.

5.3.1 A Principal Components Technique for Mapping Alteration Zones

A simple technique for alteration mapping using ATM and TM imagery, requiring no atmospheric or radiometric correction and relatively simple image processing equipment, was developed at the UKNRSC using examples of imagery of the Great Basin region of the western USA. The technique requires only a rudimentary understanding of the spectral properties of minerals and vegetation and it relies on the ability of the principal component transform to map increasing subtleties of data variance into successive components. The study was guided by detailed field information on numerous and varied mineral prospects and deposits covered by the imagery and it drew on the collective experience of exploration geologists who had been using remote sensing as a prospecting aid. The technique has been tested on ATM and TM image data acquired over some recent western US gold discoveries before or soon after their discovery. Many of these could have been located using the modified principal component analysis method, and subsequent trials of the methodology on TM images from other parts of the western United States, southern Spain, the eastern Mediterranean, the Middle East and South America showed that it has wide application in arid and semi-arid terrain.

5.3.1.1. The Technique

The principal components transformation is a multivariate statistical technique which selects uncorrelated linear combinations (eigenvector loadings) of variables in such a way that each successively extracted linear combination, or principal component, has a smaller variance. The statistical variance in multispectral images is related to the spectral response of surficial materials such as rocks, soils and vegetation and it is also influenced by the statistical dimensionality of the image data. When multispectral image channels are treated as variables and subjected to the transformation, the ordering of the principal components is influenced both by the spatial abundance of the various surficial materials and by the image statistics. The influence of scene statistics, which can be both measured and adjusted, can be used in order to 'force' the transformation to provide information on the spatial distribution and relative abundance of specific surface materials. Principal components analysis can thus be used as an image interrogation technique.

Most principal components image processing software generates, not only a series of principal component images, equal in number to the number of input bands, but also statistics of the principal components transformation. This will usually include the eigenvector loadings, which indicate the contribution of each band of the input data to each one of the principal components. If the crude spectral characteristics of a particular surface type are known, in terms of which bands are likely to have a high reflectance for this surface and which low, then the principal component mapping this surface type can usually be identified by study of the eigenvector loadings. Careful selection of input bands can limit or direct the principal component images that will be generated. In all cases, the first principal component maps the generalised albedo of the scene, the feature with the greatest variance. The second principal component will often map vegetation, although this depends on the areal importance of vegetation in the scene. If not vegetation, the second component image will be the difference between the visible and infrared bands for this scene. The lower order principal components will map more subtle features such as those associated with alteration zones.

The Loughlin technique is illustrated with reference to a TM subscene of Roberts Mountain, Eureka County, Nevada. This area includes two producing gold mines, Gold Bar and Gold Bar extension, and a number of other prospects. Gold mineralisation is epithermal in type, of probable early Tertiary age, and is hosted in Palaeozoic carbonate rocks. Mineralisation is controlled by NW - SE trending structures within a similarly trending major mineral belt known as the Battle Mountain - Eureka trend, itself parallel to the famous Carlin trend. The locations of mines and prospects are shown in Figure 5.1, a first principal component (albedo) image of the study area.

Figure 5.1. The Roberts Mountain area, Eureka County, Nevada. Gold mines and prospects are indicated. This is a first principal component (albedo) image derived from TM.

Lengthy experimentation indicated that if only four bands of the possible six TM reflected wavelengths were used in the principal components transform, and if these bands were selected so as to emphasise the desired feature, unambiguous mapping of selected types of alteration zones could be acheived. The two types of alteration readily apparent in TM imagery are "hydroxyl" alteration, highlighted by the presence of hydroxyl-bearing phyllosilicates which cause strong absorbtion in TM band 7, and limonitic iron oxide alteration, which causes absorbtion in bands 1 and 2 and higher reflectance in band 3. In order to map hydroxyl alteration, input bands are restricted to TM 1, 4, 5 and 7, omitting bands 2 and 3 in order to avoid mapping iron oxides. Reference to the eigenvector loadings allows the identification of the correct principal component, and also permits correction of the principal component image, in that the feature required can be mapped either as positive or negative, depending on the relative abundance in the scene. An example of a "hydroxyl" alteration image generated by this process is shown in Figure 5.2. Areas of alteration show up as bright patches on the image. Iron oxides are mapped by principal components analysis of TM bands 1, 3, 4 and 5, band 7 being omitted to avoid hydroxyl mapping. The "iron oxide" image will usually be the fourth principal component, but this can be checked once again by reference to the eigenvector loadings. An "iron oxide" image of the study area is shown in Figure 5.3, where bright patches identify iron oxide alteration.

Figure 5.2. "Hydroxyl" image of the Roberts Mountain area, derived by directed principal component analysis of bands 1, 4, 5 and 7 of TM. Areas of hydroxyl alteration appear bright.

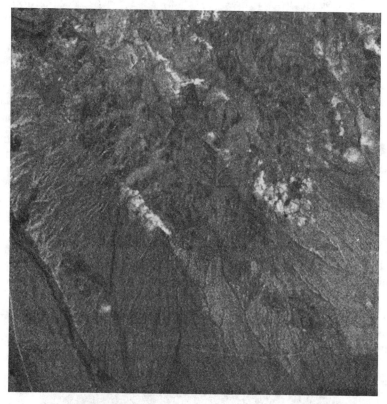

Figure 5.3. "Iron oxide" image of the Roberts Mountain area, from directed principal components analysis of bands 1, 2, 4 and 5 of TM. Areas of iron-oxide alteration appear bright.

These monochrome images are useful in themselves, but it would be preferable for interpretive purposes to combine images of both types of alteration in a colour composite. Plate 10 illustrates one possible colour composite of the Roberts Mountain area, with "hydroxyl" alteration displayed in red, "iron oxide" alteration in blue and an image summing the two principal components derived alteration images as green. Where both types of alteration coincide, the colour composite will appear white. A comparison with Figure 5.1 shows that the locations of the mines and prospects have been highlighted, and that other areas of alteration worthy of further investigation on the ground are also present. It should be noted that the image extract used in the test was acquired in 1985, before production had started from either of the gold mines in the area.

The technique has been tested on Daedalus scanner and TM images from other mineralised areas in arid and semi-arid environments, and has been shown to give consistent results.

5.4 THE IMPORTANCE OF DATA INTEGRATION

Remote sensing can provide, under ideal circumstances, three types of information of great importance in mineral exploration; structural, lithostratigraphic and mineralogical. In the worst case of dense vegetation, thick transported soil cover and subdued topography, none of these are possible and remote sensing in the sense covered in this book can play no part. Usually, however, some valuable structural information can be derived from satellite or airborne digital imagery. Under favourable conditions, usually but not always in arid environments, much information can be gained on the spatial distribution of significant rock units. Under certain circumstances, and with some types of mineral deposit, the distinctive mineralogy of alteration zones associated with mineral deposits can be discriminated using remote sensing. In a few cases, the geochemical halo associated with mineralisation may be sufficiently toxic to produce vegetation changes, and remote sensing geobotany may be possible. It is very rare, however that remote sensing can, unaided, result in the discovery of significant mineral deposits. Most of the obvious mineral deposits discoverable by their distinctive surface expression have already been found even in remote parts of the world. The energy and thoroughness of prospectors through the ages should never be underestimated. Even though the deposit may not have been mined on a large scale or recorded, it is rare to find a large mineral deposit (of materials usable before this technological age) with any surface expression whatever which does not eventually reveal traces of previous interest. Undiscovered deposits are more subtle, and their recognition demands the careful combination and analysis of data from a wide range of sources. Structural information from remote sensing and surface mapping; lithostratigraphic data from regional mapping, supplemented possibly by remote sensing; geochemical data from stream and soil samples, supplemented possibly by remote observation of geobotanical or mineralogical haloes; geophysical information from airborne and ground surveys; the results of all previous exploration in the study area, all need to be combined in such a way that the experienced explorationist can observe the inter-relationships of the multiple parameters and then use his judgement to decide on the next step in the exploration programme.

As discussed before, the term geographic information system can simply be a long name for a map. A series of transparent overlays, drawn to the same map projection and scale, can be used to study the spatial relationships between different data sets. Such analogue GIS's are limited to qualitative analysis at fixed scales, and can rarely be used for simultaneous inspection of the inter-relationships of more than three data sets. Digital GIS's do not have these intrinsic limitations. The number and size of data sets is limited only by the storage and processing power available, and by the cost of acquiring and digitising the necessary data, but even these limitations are becoming less significant each year as computers become more powerful as well as cheaper, and the amount of data acquired and stored in digital form expands. Plate 11 shows the type of product that can result from data integration, combining filtered satellite imagery to highlight structure, a digitised geological map with locations of mineral deposits, and a digital gravity map. The image is presented as a stereo pair, with apparent elevation controlled by gravity.

For the mineral explorationist, GIS's can simply be a tool to enable him to make more efficient use of the data that he acquires, either during a mineral exploration programme, from public domain sources or from other companies. In the competitive business of mineral exploration, intelligent use of GIS techniques can give a progressive company a vital edge in time and in detecting subtle inter-relationships between favourable indirect indicators. Most successful exploration companies have been in the data integration business for decades, and few recent discoveries have resulted from interpretation of a single data set in isolation. The difference that digital processing brings is one of scale rather than technique, and few companies can afford to forgo the advantages that GIS's can bring. Remotely sensed data often forms an essential component of mineral exploration GIS's because its digital nature lends it readily to integration with other data sets, because it can provide valuable indirect indications of the presence of mineralisation complementary to information from geochemistry and geophysics, and also because remotely sensed imagery can provide a recognisable geographic background against which to examine other data sets.

As an example of the sort of data integration that can assist in mineral exploration, a study was carried out at the United Kingdom National Remote Sensing Centre. This was a cooperative project between Tony Harding of the NRSC and Mike Forest of the British Geological Survey, and I am indebted to them for their permission to use the example here.

5.4.1 An Example of Data Integration

The study used a raster-based image processing system to combine geochemical, geophysical and geological map information with structural indications from remote sensing, and then used image processing techniques to model the relationships between the various parameters and interpret these in the context of standard mineral deposit occurrence models. The British Geological Survey maintains a digital geochemical database for much of the British Isles, based on stream sediment samples. A portion of this database covering the English Lake District was made available for the study. The BGS also has a gravity map of Britain in digital form, and a portion of this was extracted for the study and an analogue total magnetic field map digitised. The BGS has digitised its 1:250,000 scale geological maps, and the portion covering the Lake District was extracted. A portion of a Landsat TM image covering the study area was geometrically corrected for co-registration with the other data sets.

Although rock outcrops are numerous in the Lake District, the steep slopes and abundance of surface encrustations of lichens, mosses and other vegetation precludes the extraction of lithological or alteration information from remotely sensed imagery. The same rugged topography results, however, in the digital imagery being a valuable source of structural information. Digital filtering of the imagery was used to enhance major lineaments at an early stage in the study, and later lineament extraction techniques were used for automated detection of linear features as an indicator of fracture density.

Stream sediment geochemical data can, with little or no sophisticated processing, highlight geochemically prominent mineral deposits, but much complex processing may be necessary to detect more subtle anomalies. One useful approach is to subdivide the geochemical sample sites according to the lithological units which they represent, and then to derive threshold values for each unit. The background value for copper, for example, in basic volcanics is likely to be much higher than in quartzites, and a value not significantly above the mean for volcanics could be highly anomalous if found in quartzites. The digital geological map was used to mask the geochemical sample sites, and thus to subdivide them according to lithology. Thresholds for economically interesting metals were derived for each major lithological unit, and applied statistically to the data to identify lithologically normalised geochemical anomalies.

In the absence of a reliable geological map, the aeromagnetic maps would have given useful information on the boundaries of volcanic horizons, but the data was noisy and software available at the time did not permit rapid conversion of digitised contours to a raster magnetic field image. No detailed use was therefore made of magnetic information, although it would undoubtedly have added significantly to the study. A combination of the digital gravity map with the geological map and structurally enhanced satellite imagery was specially prepared as an interpretative base to permit visual qualitative interpretation of the regional geological structure. A dominant element as far as mineral deposit genesis is concerned is the presence of a large, mainly buried, granite mass whose emplacement appears to have been partially controlled by regional-scale fractures. The combination of gravity, surface geology and structural fabric allowed an improved interpretation of the shape of the buried granite.

Some past mining activity in the study area has exploited hydrothermal vein deposits of lead, zinc and copper, apparently emplaced in fracture systems above the buried granite. An empirical mineral exploration model for the area would therefore be to search for geochemically anomalous lead, zinc and other elements of the granitic hydrothermal association, in areas of intense fracturing with close spatial relationship to the granite surface. Geo-referencing and superimposition of significant data sets, and the production of analogue products highlighting the relationship between the various parameters, is an approach that could be used in a real mineral exploration situation. A range of composite products were prepared during this study to investigate the best methods of presenting multiple data sets to aid visual interpretation. One of the most successful of these composite products is shown in Plate 12, which combines structural elements from satellite imagery with regional gravity, geological map and geochemical anomaly information.

In order to make full use of the processing power available, especially if large areas are being screened during an exploration programme, digital processing guided by experienced explorationists should be used to highlight the most prospective areas for particular classes of mineral deposit. In this study, applying the hydrothermal vein deposit occurrence model, a measure of fracture density, proximity to buried granite, and lithologically normalised geochemical anomalies were combined to highlight areas worthy of further investigation. Digital lineament extraction techniques were used to extract lineaments from satellite imagery (Figure 5.4), and image processing used to derive images of apparent fracture density (Figure 5.5). Proximity analysis was applied to an interpretation of the shape of the buried granite derived from geological and geophysical data, and "corridors" within specified distances from the granite contact established (Figure 5.6). A digital superimposition of these corridors with areas of high fracture density defined structurally favourable sites, and the addition of coincident geochemically anomalous values allows further reduction of the area of search (Figure 5.7).

No follow-up study could be carried out to test the validity of the approach used in this demonstration study because the Lake District is a National Park with no mineral exploration licences granted, the main reason why the British Geological Survey agreed to make their data available for the demonstration. Some major mining groups are however using similar approaches, and the availability of demonstration material such as this has encouraged others to think of the possibilities of GIS approaches. It is most important that the inputs to such a system are selected by experienced explorationists, who should also define the models to be used. GIS techniques are not a "black-box" solution to exploration difficulties, but simply an expansion from traditional techniques, making possible a better appreciation of the subtle inter-relationships between different data sets, and allowing the search of larger areas in less time than by conventional approaches.

Figure 5.4 North-westerly trending lineaments in a small test area in the English Lake District, extracted from TM imagery.

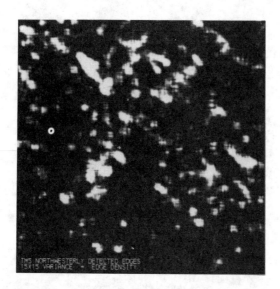

Figure 5.5 An image of density of north-westerly lineaments derived from Figure 5.4 using a 15*15 variance filter.

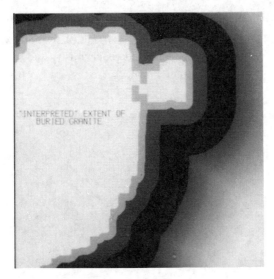

Figure 5.6 An image of proximity to the interpreted extent of a buried granite. Grey tones indicate different distances from the granite contact.

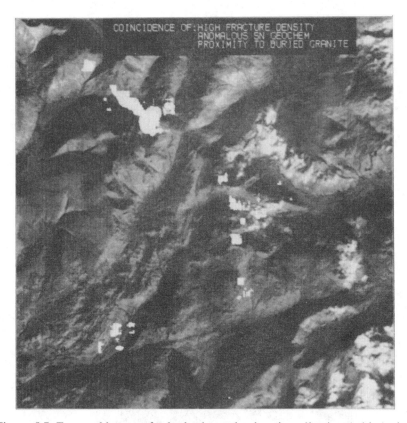

Figure 5.7 Favourable areas for hydrothermal vein mineralisation (white), derived from fracture density, lithologically corrected geochemistry and proximity to the granite contact.

6
Mine Planning

Remote sensing has been relatively little used in mine planning, but experience from other disciplines suggests that it could be a cost-effective source of additional information during the planning stage of any large mining venture, particularly in remote areas.

6.1 TRANSPORT ROUTES

In planning a new mining venture it is essential both to know the current status of transport routes in the area of the proposed mine, and to have all the information required for planning expansions and new transport networks. Initially, it must be possible to transport all the material and staff required during the development and construction phase, while later the production of the mine must be carried to markets. In the case of high-value commodities such as gold and diamonds, for example, the latter is a minor consideration, since it would be economic even to use helicopters, and no surface transportation of the mined products is really required, but in the case of lower unit value bulk commodities such as iron ore or bauxite the availability of efficient transport may be a deciding factor in the economic assessment of the new mining venture.

Satellite remote sensing can provide a low-cost and up to date source of information on pre-existing transport networks. Roads, railways and waterways are clearly visible in fine-resolution satellite imagery such as that from Landsat TM or SPOT, and in all parts of the world except those affected by persistent cloud cover imagery acquired within the past year is readily available. Maps are rarely as up to date as satellite imagery, even in developed countries, and in much of the developing world they may be as much as twenty years out of date. The amount of infrastructural development that has occurred in the past two decades in the previously less-developed parts of the world means that maps are usually very unreliable as sources of information, especially on road networks. No very sophisticated processing is required if fine-resolution satellite imagery is simply to be used as a source of information for interpretation visually by mining company cartographic staff. If no in-house image processing facility is available, then it is probably best to purchase geometrically corrected and contrast-stretched false-colour photographic hard copy at a scale of 1:50,000 or 1:100,000. This can then be interpreted in conjunction with such maps as may be available, supplemented by some field checking, in order to produce a reliable base map of transport networks. No special enhancement is usually required to detect roads, railways and rivers. The latter are usually very prominent in any imagery incorporating one or more infrared bands, since water absorbs most infrared radiation, and water bodies appear dark. Roads are often clearly visible, although their distinctiveness will depend on the contrast between the road surface and its surroundings. In a vegetated area, gravel roads often stand out more clearly than narrow tarred roads. New roads are much more distinctive than older ones, since the construction scars, borrow-pits, drains and similar features have not become masked by vegetation and trees have not yet grown along the roadsides and started to overhang the road. In some semi-arid environments, older roads may be made more visible by the settlements which have developed along them, and by trees planted by villagers along the roads. Roads are most difficult to see in densely forested areas, and it may be preferable to obtain imagery acquired at the season of minimum leaf cover, if such a season exists, to improve visibility. Railways are often less visible than roads, particularly if the railway is only single-track and narrow-guage. The type of material used as ballast is important in improving contrast, as is the amount of traffic on the line. A heavily used railway is unlikely to have much vegetation growing between the tracks, and oil and coal spills will improve visibility. A rural line with only two trains per day will probably have abundant grass growing along the track, and may be indistinguishable from the surroundings. The distinction between railways and roads can often be made on the basis of the relative uniformity of railways as compared with roads. The restrictions on curve radius and gradients which railways impose means that they are usually much more regular and geometric features than roads, and tend to follow the contours of the topography more closely. Another set of man-made linear features which are often very striking in satellite imagery, and which can occasionally be confused with roads, are powerlines. Although the actual pylons and cables are only rarely visible, the broad cut lines required where powerlines cross forested areas are very obvious, and can confuse the inexperienced interpreter, especially when a powerline access road follows the cutline.

The second major use of remote sensing at this stage of a mining venture is in planning new transport routes. The detailed planning of a road or railway will obviously involve ground surveying of the selected route, and probably also soil sampling and drilling, but the choice between different possible routes, or even modes of transport, must be made at the planning stage and can be greatly assisted by remote sensing. The integration of remote-sensing derived information on existing routes, landforms, land-use and geology, with existing topographical information and data on land ownership, population and climate in a GIS can facilitate the objective selection of the economically and environmentally preferable alternative.

As a hypothetical example let us consider a proposed new iron ore mine in a remote area of a tropical country, 150 km from the nearest suitable port site. The only access at the time of the feasibility study was along a narrow tarred road from the coast to within 30 km of the deposit, then along unpaved gravel roads built during the exploration phase. A major river flows to the coast, passing 10 km from the mine site.

Four alternative modes of transportation, used either solely or in combination, were considered by the mine planners. These were an improved road, a railway, the use of river barges, or a slurry pipeline. Topographical maps of the area, on a scale of 1:100,000, were last revised twenty years previously, and the geological teams involved in mineral exploration had found them very unreliable. It was decided to use satellite remote sensing to provide a range of information relevant to the possible options, and to use a GIS to process this information and provide assistance to the mine planners.

For both the road and rail options, the route used by the existing road and bush-tracks had some extreme gradients, and would have been unsuitable. Interpretation of geometrically corrected satellite imagery in conjunction with the old topographical maps, which had some contour information, allowed the selection of a route with better gradients, but also indicated that some substantial embankments, cuttings, and probably some tunnels would be required. Geological maps of the area were very sketchy, so Landsat TM imagery was used, in conjunction with geological information gathered during mineral exploration, and additional data collected by brief field visits. A geotechnical map was prepared for critical parts of the road or rail route, making particular use of fracture information from satellite imagery, and it became apparent that unstable rock conditions over parts of the route would require expensive engineering works. The geological interpretation was also directed at locating sources of suitable aggregate for construction along the route, and further ground checking confirmed the availability of suitable material along much of the route. A proximity analysis between the proposed route and possible aggregate sites indicated that haulage distances would not exceed 10 km, except in the coastal plain where thick alluvium and a consequent lack of solid rock outcrops meant that aggregate would have to be transported more than 30 km for this portion of the road or railway. During the feasibility study, a new national park was declared in the hills between the mine-site and the coast. Although the proposed road or rail route would not pass through this park, the government authorities requested the mining company to demonstrate that the proposed route would not be visible from any of the proposed tourist lodges in the new park. A crude digital elevation model was prepared from the existing topographic maps, and the road/rail route and proposed lodge sites draped over this digital elevation model using a GIS package. Intervisibility studies showed that only a small portion of the route would be visible, and government authorities agreed that, should the road or railway be embarked on, careful construction and the planting of trees would minimise the environmental impact. The remote sensing study, which included the comparison of recent satellite imagery with low-cost pre-1980 Landsat MSS imagery, also indicated that considerable agricultural development had occurred along the coastal plain during the previous decade, and that transport networks for agricultural produce were poorly developed. The possibility of government participation in part of the road or rail construction was suggested, since the proposed new route would have considerable collateral benefit for the new settlers. Two alternative possible routes for the road or railway through the agricultural lowlands were suggested following analysis of land-use information in a GIS. One route would serve the maximum number of people, in case the government wished to make use of the new route, and was prepared to arrange for land acquisition along it, while the other was sited to pass through less favourable agricultural land to minimise the impact on current settlement.

The possibility of a river route was also investigated with the assistance of remote sensing. The river runs in a deep valley below the mine site, and descends a total of 150 metres from immediately below the mine site to the sea. Initial examination of satellite imagery (in this case SPOT, to take advantage of the finer spatial resolution) suggested that there are major rapids immediately downstream from the mine site, and field checking confirmed that more than 70 metres of the descent occurs here, the remainder being at three other sets of smaller rapids between here and the edge of the coastal plain. Geological and topographic studies of the areas around these rapids indicated that, in two cases, the construction of weirs and locks would be feasible. The third set of rapids is enclosed in a deep valley, and construction of any canal would involve prohibitively large volumes of excavation in solid rock, but a visit to the site indicated that construction of an inclined plane for transport of barges up and down would be possible. The river meanders across the flood plain, and then enters the sea via a broad delta. Since the meandering portion of the river and the delta might be expected to be dynamic environments subject to rapid change, it was decided to carry out a brief remote sensing study of changes in the recent past, and thus to estimate rates of change and identify sites of greatest change. Old Landsat MSS imagery, acquired in 1974 and 1980, was purchased and co-registered with recent Landsat TM imagery. The three dates of co-registered imagery confirmed that some meanders are moving at detectable rates, and that there was some major realignment of river channels in the delta between 1974 and 1980. A brief visit to the meteorological office in the capital confirmed that exceptionally heavy rainfall in 1976 resulted in extensive flooding, and that the river course through the delta changed at that time. The amount and rates of change detected indicated that some engineering works would be required on the most mobile meanders, and that regular dredging would probably be required to keep a channel open for barges.

Remote sensing could not provide much information relevant to decisions on the slurry pipeline option, since it was less dependent on topography or surface conditions, but since it would be desirable to route the pipeline away from population centres and watercourses in the event of leakages, proximity analysis of a settlement and drainage map prepared by interpretation of satellite imagery was used to check possible pipeline routes.

All this information was presented to the planners, who then used this in their detailed feasibility study, and eventually decided on the best mode and route of transport for the iron ore. This hypothetical study would have required the purchase of the satellite imagery listed in Table 6.1.

The image processing and GIS work would require approximately three person-months of effort. If the company had its own image processing and GIS facility, the total cost of the remote sensing and GIS components, excluding field checking (which would have to be carried out in greater detail if remote sensing was not used) would be of the order of 35,000 US dollars. If carried out by a remote sensing contractor, which would not be an ideal approach owing to the requirement for continued interaction between planners, remote sensors, and field staff, this cost would probably be increased by 50 to 75 %.

Table 6.1. Satellite imagery required for transport routes study

Sensor	Quantity
Landsat TM	1 full scene
	3 quarter scenes
Landsat MSS	4 full scenes
SPOT XS	2 scenes

6.2 NATURAL HAZARDS

Of the wide range of natural hazards which could affect a mining operation, those which can be assessed, at least in part, by the use of remote sensing are earthquakes, landslides, floods and volcanic eruptions. These natural hazards are a danger to many aspects of mining, including the extraction itself, whether underground or open-cast, the concentrators and smelters, transport connections to the outside world, and of course the housing areas of mine employees. Short-term warning of an impending disaster can use remote sensing, in parallel with many other data sources, but it is not this application with which we are concerned in this section. For mine-planning purposes what are required are estimates of the likelihood of the occurrence of natural hazards, rather than predictions of the time of the occurrence.

The probability of occurrence of earthquakes, landslides, floods and volcanic eruptions can be deduced from the frequency of previous occurrences, interpreted in the knowledge of the mechanisms which drive and facilitate these events. Of the four categories of natural hazard, earthquakes are probably the most difficult for which to derive direct information on frequency of past occurrence. Many of the effects of even very large earthquakes are rather transitory, while other effects can be confused with different natural hazards. For example, an earthquake can trigger a landslide, which in turn blocks a river to cause a flood. The majority of serious earthquakes are associated with active fault zones, especially those at plate margins with lateral slip predominating over vertical movement. Such fault zones are usually clearly visible on satellite imagery as distinctive linear features, and evidence for lateral slip appears in displacement of lithological horizons and older fracture zones. Evidence of geologically recent movement is important, as an ancient fracture zone which is now inactive is unlikely to result in serious earthquakes, even though it may have involved huge movements in the geologically distant past. Evidence for recent movement must be sought in the displacement of geomorphological rather than lithological features. Stream courses may be displaced, spurs of ridges may be truncated, and lakes (almost always geologically transient features) may be produced along one side of the fault zone. An assessment of seismic risk can be based on the presence of major active fractures, although the density of fractures will probably be of little assistance. A single large fault zone in otherwise unfractured rock is likely to produce more serious earthquakes than a dense network of fractures, which allow more gradual non-catastrophic movement. A risk assessment based on structural studies must always be quantified by reference to historical records, if the area is inhabited, or else by comparison with geologically similar areas elsewhere in the world.

Evidence of landslides is often clearly visible in satellite imagery. The characteristic shapes of major slides, especially if relatively young and therefore producing areas of disturbed vegetation, are best seen in stereoscopic imagery, but are still visible in monoscopic examples. Interpretation must be directed at identifying different age classes of landslips, based on vegetation recovery, erosion of the toes of slips by streams, and the reduction in surface roughness that occurs as a slip ages. The number and size of slip in each age class is then assessed by detailed study of the imagery. Quantification of the actual age of the different classes should be attempted if possible, drawing from historical records and possibly also making use of absolute age determinations of wood fragments and other organic material trapped in slip debris. The frequency of occurrence of significant slips can then be calculated, and the risk assessed. The occurrence of major landslips is often closely tied to climate, in that high rainfall usually increases the risk of slips. The risk assessment should also include information on climatic trends, if available, to suggest whether the risk of landslips has increased or decreased over the time period used for the sample.

Previous occurrences of floods are difficult to assess by remote sensing. A recent flood leaves dramatic traces, including a blanket of mud and silt and a "high water mark" of flotsam, but these relicts disappear fairly rapidly, even in uninhabited areas. New sediment is rapidly covered by vegetation, and old vegetation killed by the flood decays quickly in most environments. One clue to regular flooding can be found in patterns of settlement and agriculture. Except in the most densely populated countries like Bangladesh, and in arid environments where silt carried by floods is an essential part of agriculture, settlements and intensive agriculture are rarely found in areas subject to regular flooding. Observations in satellite imagery of extensive areas of flat land along river courses without settlements, and where agriculture is restricted to pasture lands, would be a strong indication of regular flooding. Even in developed countries such as Britain this simple indication is often forgotten or ignored. "Water meadows", areas of pasture land along rivers, were never built on or cultivated intensively because of local knowledge of their being subject to flooding. Recent development of housing estates on such land has often been followed by disastrous flooding, simply because the developers ignored local folklore. In completely uninhabited areas, extensive tracts of natural grassland along river courses are a good indication of periodic flooding, since such conditions inhibit the growth of many tree species.

Evidence of past volcanic activity is normally clearly visible in satellite imagery. Volcanic landforms are very distinctive, and can provide information on the predominant type of eruption as well as its frequency. It is relatively simple to deduce from satellite imagery whether a volcano erupts predominantly basaltic lavas, with little risk of explosive pyroclastic eruptions, or whether the vulcanism is mainly acidic in character, with pyroclastics predominating over lavas, and the risk of violent eruptions greatly increased. In the case of volcanoes liable to pyroclastic flow or mudflow eruptions, which can have disastrous effects far from the source volcano, some predictions can also be made on the likely courses followed by such features, based on the morphology of the area and the sites of previous eruptions. Volcanic rocks, whether pyroclastic or lavas, decay at rates dependent on their composition and on the prevailing climate. The slopes of pyroclastic cones decrease with age as the steeper portions are eroded and slump to the natural angle of rest of the material. Slopes are difficult to measure quantitatively without stereoscopic imagery, but qualitative assessment of relative slopes is possible based on the fact that most satellite imagery is acquired at intermediate sun angles, at about 1000 hrs local time, and shadowing on steep westerly slopes will be prominent. Under most climatic and chemical conditions, lava and pyroclastics will gradually decompose and become covered with vegetation. Some volcanic rocks either decompose very slowly or else are toxic to vegetation, so that the recovery is extremely slow, while others weather rapidly into mineral-rich soils which encourage lush vegetation growth. For any volcanic district, a local time scale must be established in order to permit reasonably accurate estimates of eruption frequency by quantifying the age classes determined by remote sensing.

6.3 POPULATION DISTRIBUTION

It may come as a surprise to many that it is possible to estimate population using satellite imagery. People are much too small a target individually for even the finest resolution commercial satellite imagery, and it is rarely even possible to count separate houses unless they are very large. Rough population estimates can however be made using remote sensing. These are based on settlement sizes, the areas of urban development and the amount of land under cultivation. Calibration for different cultural and climatic settings is essential, but results markedly more reliable and up to date than many official census statistics can be obtained under favourable circumstances.

Planners of a new mine may need information on population density and distribution for a variety of reasons, but there are three main uses. Firstly, the planners need to be able incorporate population information in their environmental impact assessment. The effect of a new mine on people resident in the area and not directly connected with the mining operation should obviously be considered, and the total number of people affected must also be known. Their socio-economic status may also be important. While, in an ideal world, the impact of new development on people should be assessed as an absolute, independent of the status or influence of the people affected, mining companies and government agencies will take most notice of the people who are likely to give them the most trouble. The location of residential and recreational areas of the upper socio-economic groups relative to a proposed mine site is likely to be a matter of special concern to the planners. A second reason for requiring population information is the need for labour. If there is a large resident population in the general area of the mine, this may serve as a source of labour, removing the need to construct expensive housing areas and related infrastructure, although the higher level of technological skills required in many modern mines reduces the importance of low-cost unskilled labour. A third reason for requiring population information is the availability of housing for mine staff. If there are already towns, villages and roads in the area of the proposed mine, it may not be necessary to construct a completely new mine township, but the company may rather choose to assist with improvements to existing facilities and infrastructure.

Information on population density and distribution is either lacking or unreliable in many countries, developed as well as developing. Even where a regular census is conducted, the results may be deliberately distorted for political or other reasons. Local government officials may inflate population statistics for some areas in order to attract additional central government funding, or to justify extra parliamentary constituencies. Conversely, census figures may be deliberately reduced in order to cover up discrepancies in taxation which should have been passed on to central government. The rate of population growth and migration in many countries is such that, even with reliable censuses conducted every five years, the figures rapidly become outdated. Satellite remote sensing can provide approximate figures on total population, and more precise information on the distribution of population.

Population estimates are based on extrapolation of detailed ground samples to progressively larger areas. The procedure would be firstly to examine satellite imagery of the area of interest with particular reference to settlements. A preliminary classification of settlement by size and activity (village, small town, large town, group of farms, market town, industrial town) should be attempted at this stage, and representative examples of each category selected for more detailed checking. If recent air photos are available for the study area, the number of houses per settlement, the number of houses per unit area of agricultural land in farming communities and the number of houses per unit area of high-density and low-density urban areas should be counted for a reasonable number of test sites. Information is then required on the average number of people per household in each of the broad socio-economic categories. This may sometimes have to be collected by field visits, but may be available from governmental or academic sources. A series of equations can then be established, relating population to areas of cultivated land and to areas of recognisable classes of urban land. In some cases factors such as the density of rural tracks and the degree of encroachment into forest lands may provide additional clues, and the interpreter may have to apply some lateral thinking to discover other clearly visible signs of human occupancy. The whole study area can then be classified by a combination of automated processing (for areas of agricultural land and some urban areas) and visual interpretation (for areas and numbers of settlements), and a population map prepared. The results are obviously not to be treated as an accurate census, but a surprisingly good estimate of population density and distribution can be prepared rapidly and relatively cheaply by this means.

The identification of residential and recreational areas of the higher socio-economic groups is surprisingly easy using remote sensing. The signs of affluence are apparent even from a distance of a thousand kilometres. Swimming pools, broad lawns, tree-lined streets, golf-courses, ski-runs and airfields, for example, are all very clearly visible in TM or SPOT imagery, and the planners can pin-point, at a very early stage of the study, those critical areas which are likely to be most vociferously concerned by the mine development. In some cases, the total absence of population and lack of signs of human interference are not necessarily good news for the planners and developers. Such areas, being unspoilt, may already have been declared as wilderness areas, or the open interest of a mining company may encourage people to urge their preservation.

6.4 FORESTRY AND AGRICULTURE

Apart from the importance of mapping existing land use around a new mine site as an input to the environmental impact assessment, discussed in Chapter 7, mine planners may require information on agriculture and forestry for other reasons as an input to their economic analysis.

In the case of a large new mine employing many people, local sources of basic foodstuffs will significantly reduce the operating costs, both in the direct costs of importing these goods, and in the management overheads required to organise import and distribution. Information on the area of different crops and types of agriculture in the region of a proposed new mine could serve as an indicator of the types and even quantities of produce that are locally available. Provided that some ground data on agricultural practices are available, classification, especially of multi-date TM imagery, can provide a reliable indication of the main categories and areas of agriculture in the region. GIS techniques could incorporate this information with existing and planned transport networks, and with settlement patterns and densities, and thus assist in economic analysis.

The availability of natural and planted timber may be important in some underground mining operations where extensive support is needed. The use of wood as a fuel in smelting is decreasing due to environmental pressures on the mining industry, but this energy source may still be considered in some cases. Satellite remote sensing is especially well suited to forest mapping, and the mid-infrared bands of Landsat TM permit subdivision of woodlands into a number of environmentally and economically important categories. In many tropical and sub-tropical areas, plantations of conifers and eucalypts can be clearly distinguished from natural forest, while the latter can also be subdivided according to photosynthetic activity and density. A single date of cloud-free TM imagery is usually sufficient to prepare a forest map of a very large area, although some field checking will be required to "fine-tune" the classification process. Precise estimates of the volume of standing timber, or even of the mean tree height, are rarely possible, however, except in very uniform plantations, and such plantatations will normally keep accurate records, obviating the need for remotely sensed information. Imagery from synthetic aperture radar sensors is very useful in forest studies, since differences in leaf shape and structure strongly influence the amount of radar backscatter. Such imagery may also be more readily available in equatorial areas than optical imagery, due to problems of cloud cover. Automated processing of this imagery is less well developed than for optical sensors, depending as it does on texture more than on perceived reflectances, but experience with ERS-1 is likely to improve the classification accuracies attainable with SAR.

6.5 SELECTION OF SITES FOR HOUSING AND DUMPS

In a new mining venture in relatively uninhabited country, the selection of a suitable site for staff housing can be critically important to the success of the venture. In a remote and isolated community, staff morale must be promoted by all practical means, and comfortable and relaxing surroundings can play a major part in this. Remote sensing can contribute to the successful selection of a housing site.

An ideal mine housing area should not be too far from the actual mine in order to minimise costs and time taken for travelling, but residents should not be conscious of their proximity to the mine. The site should be scenically interesting where possible, taking advantage of any topography to provide varied views as well as the possibility of breezes in tropical environments or shelter in less temperate climates. There should be sufficient flat land for recreational purposes (football pitch, swimming pool, etc) and slopes should not be so steep as to risk landslides or drastically increase construction costs. The site should not be liable to flooding, and should be accessible at all seasons. In moister climates, existing tree cover should be retained where possible to provide shade and interest, while in arid zones the possibility of using waste water from the township or even the concentrator to promote a greener environment should be investigated. Disposal of waste, both waterborne sewage and dry household waste, is important and suitable environmentally benign sites must be chosen for this. All these points may seem self-evident, and the use of remote sensing in site selection may not be obvious.

In isolated and less-developed areas, remote sensing can be used in an initial selection of possible sites, based on vegetation and topographical diversity as well as proximity to the mine site. Once some possible sites have been short-listed, the choice between them can be quantified using combinations of remote sensing and GIS. The topography can be modelled in digital form, based on contour plans of possible sites produced by the mining company, or even derived from stereoscopic satellite imagery, and studies can be done of the visibility of the mine site and proposed dumps from the possible housing areas. Combinations of satellite imagery and digital elevation models can be used to prepare perspective views of different possible sites to aid the decision making process. Vegetation maps prepared from remote sensing can provide quantitative information on the types, diversity and spatial distribution of different cover types which can be incorporated into a GIS. Some information on possible natural hazards which could affect the housing area could be derived from remote sensing. Old landslip scars, evidence of flooding, and locations of active fracture zones could all be derived from satellite imagery. The relative location of areas of surface water, streams, ponds, lakes and marshes, to the housing area could be important from a recreational angle and for reasons of health as well as for water supply and waste disposal. In many tropical countries mosquitoes and other insect pests depend on surface water for breeding, and while careful control of the immediate surroundings of housing areas by the mining company can reduce the number of potential breeding sites, the effects of proximity to a major swamp could not be so easily reduced.

As an example, we could consider the use of remote sensing and GIS techniques in the selection of housing areas for a major new mine site in a remote mountainous semi-arid environment. Initial remote sensing studies have located three possible sites for the housing area within ten kilometres radius of the mine site. These sites were selected on the basis of their elevated topographical positions and apparent scenic merit. For each potential site a topographical map is prepared using a combination of old air photography and recent SPOT imagery. Land-cover and geological information is derived from Landsat TM imagery, with some ground checking during a short field visit by the planning team. A geographic information system is prepared for each site containing the following layers:-

a) Topography in the form of a digital elevation model.
b) Land cover, including distribution of natural vegetation classes (scrub, grassland, non-vegetated), main soil units (sand, gravel, bare rock) and indications of active erosion.
c) Drainage, derived from combined interpretation of stereoscopic SPOT imagery and Landsat TM, showing main categories of drainage courses and locations of apparently permanent and transient pools.
d) Geology, including main lithological units and fracture zones.

In addition, a coarser-resolution digital elevation model of the whole mining area, including mine site and all three proposed housing sites, is prepared for intervisibility studies.

The detailed digital elevation models are then further processed to generate slope and aspect maps. Each layer of the GIS is then classified on the basis of favourable and unfavourable factors for the location of the housing area. Slopes in excess of twenty degrees are considered unsuitable for housing, as are extensive areas with slopes of less than five degrees. Since this mine is in the southern hemisphere, southerly aspects are considered more favourable than northerly to provide increased shade. Areas of existing natural vegetation are more favourable housing locations, while large expanses of bare rock are unfavourable due to the high construction costs and difficulties for gardens and landscaping. Areas with apparent high erosion rates are also to be avoided, while sand and gravel on moderate slopes is a favourable factor. The major negative geological factor is the presence in two of the proposed housing areas of a folded horizon of low-grade ore-shale, insufficiently mineralised to be mined, but with significant concentrations of lead, zinc, cadmium and arsenic. This horizon gives rise to metal-enriched residual soils, toxic to most plants, and can be readily mapped by the almost vegetation-free nature of its outcrop. An equation is generated within the GIS to give due weight to all these positive and negative factors, and a new derivative map generated which quantifies the relative favourability of each 100*100 metre cell of each potential housing area for settlement. The three sites are then compared on the basis of the total area suitable in each site, and also on the spatial relationships of individual favourable zones to each other. From a scenic point of view, a large rectangular area of favourable land, large enough to accommodate the entire housing for the mine, would be less attractive than numerous elongate patches of suitable land. At this stage the locations of major drainage courses in relation to favourable housing sites must also be assessed, since these could lead to communications problems following major rain storms. Evaluation of the coarse resolution digital elevation model of the entire mine area reveals that portions of each of the three potential housing sites would have unobstructed views of the mine site, and reference to climatic records suggests that one entire possible site would be down-wind of the mine for seven months of each year.

The results of all these analyses are then presented in a quantitative form to the planners, enabling them to make an objective selection of the most suitable housing site in terms of a combination of construction costs and the amenities of the site.

The siting of major rock dumps and tailings dams is another important planning consideration that can benefit from information derived from remote sensing and processed in a GIS. Apart from the obvious practical constraints of wishing to place rock dumps as close to the mine as possible to minimise transport costs, and of avoiding areas of future caving above underground mining operations as sites for tailings dams, there are other financial and environmental factors to be considered in the location of rock dumps and tailings dams.

With rock dumps, the major environmental consideration is likely to be an aesthetic one. Rock dumps are not usually scenically attractive, and they may need to be sited where they are at least partly hidden from view by the topography. A combination of proximity and intervisibility analysis can be used to select the least obtrusive site within the constraints imposed by haulage costs. Tailings dams pose additional problems to the planners. They need to be kept as unobtrusive as possible, but must also be sited so that leakage into surface streams or major aquifers is minimised. Proximity analysis can be used to locate areas as remote as possible from surface streams, and remote sensing can provide information on the distribution of particularly sensitive habitats in wetland areas which might be at risk should slurry escape from the dam.

7
Environmental Impact Assessment

Environmental impact assessments are becoming increasingly important in most countries of the world, and in many cases are required by law before permission can be granted for any new mining or industrial development. Most mining companies realise the public relations value of being seen to be environmentally conscious, and undertake environmental impact assessments whether or not these are required by law. With its capability for cost-effective mapping of vegetation and other surface types, and the possibilities for digital analysis of the resulting maps, remote sensing can be of great assistance in environmental impact assessments, especially where the areas to be studied are large.

7.1 MAPPING EXISTING LAND USE

One of the essentials in assessment of the environmental impact of a proposed new mining operation, or even an extension to an existing mine, is accurate information about existing land use and land cover in the surrounding area. Remote sensing is a cost-effective technique for acquiring this information at a scale appropriate to most such studies, although some field studies are always essential, and not all categories of land use are uniquely identifiable using remote sensing. A knowledge of what is and what is not possible should precede the planning of any environmental impact assessment using remote sensing.

Certain land cover types can be uniquely identified using a single date of satellite imagery, while others require different carefully selected dates, and yet others cannot be distinguished using current sensors. In order to be distinguished, and thus mapped, a surface must have either a distinctive spectral signature, enabling its identification using spectral classifiers, or distinctive texture, allowing identification by visual interpretation or by means of textural classifiers. Water bodies, lakes, ponds, dams, rivers and the open sea, can be accurately mapped in satellite imagery of any season if infrared data, whether from the near or mid-infrared (SPOT band 3, MSS band 4, TM bands 4, 5 and 7, for example) is available. The only restriction on this is that winter imagery of areas subject to snow and sub-zero temperatures is of little use, since water will be ice-covered and may be difficult to distinguish, even using textural techniques, from snow-covered land. Woodland can be distinguished from arable land, pasture and most other land-cover types in any imagery having mid-infrared bands

Table 7.1. Land-cover categories mapable using remote sensing

Category	Sensor	Notes
1) Non-vegetated areas		
Water	Any sensor with infrared	Water is the easiest
		category to
Ice and snow	and visible	map. Distinction between rock
Bare rock, soil and sand	TM (SPOT, IRS)	and soil groups is best with TM
2) Vegetated areas		
Forest		
Natural forest. Many classes depending on region		Mapping of forest types,both
Rainforest	TM (SPOT, IRS,	natural and planted, is greatly
Mangrove	ERS-1)	improved using information from
Bamboo		the mid-infrared portion of the
Savannah woodland		spectrum (TM band 5).Visual
Thorn scrub		interpretation of imagery without
Deciduous forest		this waveband (e.g. SPOT) can
Coniferous forest		give good results.
Planted Forest. Many classes depending on region		
Conifers	TM (SPOT, IRS,	ERS-1 SAR imagery is especially
Eucalypts	ERS-1)	useful in planted forest.
Rubber		
Other Natural Vegetation		
Grassland	TM (SPOT, IRS)	Many natural vegetation groups
Marsh and swamp		grade into each other, and require
Moorland		special techniques for mapping.
Cropland		
Cereals	TM (SPOT, IRS)	The separation of different crops
Wheat and barley		within the major categories of
Maize		cereals and root crops usually
Sorghum		requires multi-date imagery
Rice		within a single growing season.
Root Crops	TM (SPOT, IRS)	This can be a problem in areas
Potatoes		subject to cloud, and is also
Beets		expensive.
Other crops	TM (SPOT, IRS)	
Sugar		Some crops, especially those
Oilseed Rape		grown as monocultures over
Tea		large areas, are very distinctive,
Pasture		and easy to map.
3) Other man-made features		
Urban areas	SPOT PAN (TM)	Anthropogenic features are often
Roads, railways		identifiable by their geometry
Airports		rather than spectral signature and
Mines		thus require fine resolution.

(equivalent of TM band 5) acquired between early spring and mid-autumn. Heather (calluna and erica) can be clearly identified in all except winter imagery, as long as information from the mid infrared is available, although mapping of heather is complicated by the fact that it rarely occurs alone, but in admixtures with other moorland and heathland species. The texture of urban areas makes them readily identifiable in imagery of almost any wavelength and season, provided that the spatial resolution is finer than about fifty metres. Subtle differences in spectral signature are rarely detectable using satellite sensors in operational use at present, and since it is unlikely that sensors with significantly finer spectral resolution for vegetation mapping purposes will be operational before 1997 at the earliest, I will restrict myself to discussion of what is possible at present. Although many crops and natural vegetation communities do not have distinctive spectral signatures, they may have distinctive growth cycles, maturing, flowering and senescing at different seasons, which enable their identification in imagery of specific dates. Distinctively flowering crops such as oilseed rape are clearly identifiable at visible wavelengths during the flowering season, while cereal crops such as wheat and barley are distinguishable from grassland and pasture, which belong to the same family and thus have similar spectral signature, by the fact that the cereals ripen and are harvested at specific seasons.

Table 7.1 summarises the land-cover types which can be reliably distinguished and mapped using operational earth observation satellites. In many areas, a sufficiently accurate land-cover map of the area around a proposed new mine can be prepared using a single date of satellite imagery acquired between spring and mid-autumn, image interpretation being supplemented by field work whose duration will depend on the complexity of the area and on the critical land-cover types present. Prior to attempting any land-cover mapping using remote sensing, it is essential to identify the critical environmental cover types within the study area, and to decide whether these are amenable to mapping by remote sensing. Areas of ancient deciduous woodland, marshland or even coastal sand dunes may be the ecologically most important land-cover types in the area, and can be readily mapped. If, on the other hand, the most important features are the distribution of a particular shrub within hedgerows, or buried archaeological sites, or even some buildings of special architectural or historic interest, there is little that remote sensing can contribute. Having identified the ecologically critical land-cover types, and decided that they are amenable to mapping using remote sensing, the next stage is to establish an image classification scheme which will most effectively map the selected land-cover types, as well as other important features such as rivers, settlements and major roads, which will serve as reference points during the study. Single-date or multi-date imagery will be used as required by the particular circumstances, and it is likely that a study such as this will be carried out using imagery with information in the mid-infrared portion of the spectrum, which at present means Landsat TM. If certain important cover types are best discriminated on the basis of texture, it may be beneficial to use imagery with fine spatial resolution, at present SPOT PAN, co-registered with the TM imagery, to increase the spatial information content.

Once a land-cover map has been prepared on the basis of remote sensing and some field studies, this map must then be combined with other important map information not directly obtainable by remote sensing. The most important of these is topography, either in the form of a digital elevation model, or of a contour map with additional information such as roads, railways, towns etc. Land ownership information may also be important at this stage, since the mining company may have different environmental responsibilities within its own land from those it has in respect to its neighbours. Soil maps can sometime be prepared using remote sensing, particularly in more arid areas, but if these already exist it may be necessary to digitise them and co-register them with land-cover maps at this stage. Climatic information, not necessary in map form, but including rainfall records, prevailing wind directions at different seasons, and even information on areas subject to fog or frost, may also be relevant to the environmental impact assessment. The future plans of other organisations active in the area, whether other mining companies and industrial concerns or local and national government, should also be known and incorporated into the overall geo-referenced database. Once all the relevant information has been assembled in a geo-referenced form, analysis can be carried out.

The data for the EIA can be processed either on the image processing system used for analysis of the remotely sensed data, or on a vector GIS. Both have their advantages and disadvantages, as outlined in Chapter 3, but the final choice will probably depend on what is available to the mining company or their consultants.

7.2 ANALYSIS OF PROXIMITY TO SENSITIVE AREAS AND INTERVISIBILITY.

Proximity analysis has already been discussed in Chapter 2 (Image Processing) and examples of the use of this technique in mineral exploration and mine planning have been presented in Chapters 5 and 6. Environmental impact assessments will often require proximity analysis, since the distance of mine sites from environmentally sensitive areas is commonly a matter of prime concern.

Intervisibility studies use combinations of satellite imagery or thematic maps derived either from the imagery or other sources, with digital elevation models. Many image processing and geographic information systems allow intervisibility analysis. The process will either allow simple "yes/no" answers to whether one point is visible from another point, or will allow the generation of new maps or images showing the area visible from a single point. The converse is obviously true, in that the new map also shows all the areas from which the selected point is visible. This is illustrated in Figure 7.1. The image on the left is a TM band 3 (visible) image of part of a large china clay mine in southwest Britain. A viewpoint on the top of a large waste dump is indicated at the upper right of the image. The right-hand image shows all areas which can be seen from that viewpoint, coloured black. Most intervisibility analysis software allows interactive adjustment of the height of the "viewpoint", so that the visual impact of different heights of dumps, chimneys or headgear, for example, can be assessed.

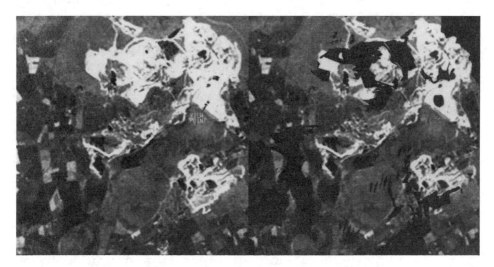

Figure 7.1. An example of intervisibility analysis. The image on the right shows areas shaded black which are visible from the top of a large mine dump at the centre right of the left-hand image. Lee Moor china clay works, Devon.

7.3 PERSPECTIVE MODELLING

One of the most visually impressive uses, to the layman, to which satellite imagery can be put is in the generation of perspective or three-dimensional views. The digital nature of the imagery and the increasing availability of digital elevation models, at least in the better mapped areas of the world, makes the generation of such views relatively simple on most modern image processing systems. Some computers are becoming fast enough to generate such views in almost real time, opening up the possibility of interactive movement of the view point and angle, and the generation of "fly-by" type moving simulations. The results of all this are very spectacular, but they are also extremely useful in a variety of situations. Mine planners often construct solid three-dimensional models of proposed new mine developments, in order to give a visual impression of the appearance of the planned developments in a realistic setting. The problem with these models is that they are very expensive to produce, and only illustrate a single possible planning alternative. They also usually only include a relatively small area of the surrounding countryside for reasons of cost, size and scale. A computer model does not have these limitations, except that the "closeness" with which one can approach any feature in the model is limited by the spatial resolution of the data used. Remotely sensed imagery is best suited to examination of the gross appearance of a mine within its setting, from any viewpoint and distance, rather than examination of the details of a pit or dump. Unmodified satellite imagery combined with a digital elevation model of the same date provides a view of the mining operation as it actually appears, but it is also possible to simulate the affects of different planning decisions by modifying both the image and the elevation model. This is illustrated in Plate 13 which shows the Lee Moor china clay workings in south Devon as they actually were in 1984, viewed from the south-west at an elevation of 1000 metres. Three bands of a Landsat TM image have been warped using a composite digital elevation model prepared from mine plans and a local topographic map. This is an isometric view rather than true perspective. Plate 14 shows the appearance of a possible hypothetical mining plan, with most of the major pits becoming amalgamated into a single large pit, and a new very large waste dump. The same combination of digital elevation model and imagery could be used to determine the areas from which this new dump could be visible. A simple iterative process could determine the maximum height of dump permissible within pre-set visibility constraints.

The generation of three-dimensional views from combinations of satellite imagery with digital elevation models has significant advantages over the more conventional computer-graphics use of "wire models" ore "shaded relief models". The actual appearance of the ground surface is more realistically portrayed, and the combination of other image processing operations with model generation allows more flexible interactive analysis of the inter-relationships of different land-cover types and mining activities. The limitations of remote sensing in this type of application become apparent if one tries to obtain too close or detailed a view of the mining operation, or of any other surface features. Neither the imagery nor the digital elevation models are sufficiently detailed for very close examination, but their strength lies in the capabilities for portrayal of the mine or other feature of interest in the context of its surroundings. Views of this type can now be produced fairly rapidly even on relatively small PC-based image processing systems, and no longer require the huge computing power that was once necessary. Digital elevation models do not exist yet for many areas of the world, although most new mapping which is carried out is compiled in digital form, requiring only fairly simple processing to generate digital elevation models from digital contour maps. For large areas, it is cost-effective to generate coarse-resolution digital elevation models from stereoscopic satellite imagery. The production of digital elevation models by manual digitisation of contour maps is tedious and time-consuming, but is a practical option for relatively small areas. The digital elevation model used in generating the Lee Moor images was prepared by scanning contour maps on a small-format colour scanner. The resulting three-band digital image was then processed on a standard image processing system to first extract contour lines from other information using a standard supervised pixel classifier. The binary contour image was then interactively edited to fill in gaps and thin lines in areas of closely spaced contours. The areas between contours were then infilled with digital values intermediate between contour values, and the resulting "layer-cake" image smoothed to give an approximate elevation model.

The use of perspective views in the mineral industry is not restricted to the examination of mining sites. The choice between alternative transport routes may often be assisted by the chance to view these alternatives in three dimensions in a natural-appearing setting. The solution of complex geological problems can be helped by a three-dimensional representation of the surface with superimposed geological information. Plate 15 shows a perspective view of the Isle of Wight, off the south coast of Britain. This image is actually a composite of three different data sets. A geological map of the area provides the colour, the surface texture is derived from a filtered Landsat TM scene, and elevation information comes from a digital elevation model. The three-dimensional view illustrates very clearly the monoclinal structure of the island, with steep northerly dips of Cretaceous beds along the centre of the island and a southward flattening resulting in repeated outcrops of isolated remnants to the south. A competent draftsman might take days to produce a single such view, while an image processing system could produce an infinite number of views of the same data set at the rate of one every few minutes once the original data had been loaded onto the system.

8
Water Supply and Waste Disposal

Water supply and waste disposal are very closely-linked topics, since the same aquifers or fracture systems which serve as sources of water can also permit percolation of pollutants from waste disposal sites such as tailings dams. Similar techniques are used to study the geology in both cases. Mining operations in remote areas usually require their own water supply, and in arid environments an adequate supply of good water for extractive processing and human consumption can be a limiting factor on mine development. Satellite remote sensing has been used for fracture analysis in hydrogeology since the days of the first Landsat, ERTS-1.

8.1 FRACTURE ANALYSIS AND HYDROGEOLOGY
The techniques used in locating suitable sites for water supply boreholes vary greatly with the geology and with the amount of water required. In areas of thick alluvium the precise location of boreholes may depend more on the positions of existing surface features (buildings, mine workings and so on) than on geology, and there is usually little surface expression of underlying lenses of water-saturated coarse alluvium. A similar situation exists where water is to be obtained from permeable horizons in flat-lying or gently folded sedimentary rocks. In areas of crystalline rocks, on the other hand, the distribution of groundwater is controlled primarily by fracture systems in the rocks, and these fractures are particularly amenable to location by remote sensing techniques. In such crystalline rocks, hydrogeological studies siting boreholes for water supply and for assessment of leakage from potential waste disposal sites usually include the detection of zones of fracturing in sub-outcropping rocks. The significance of the fracture zones, whether as secondary aquifers or as potential channels for leakage of waste, is then assessed in relation to the underlying lithology and topographical position of the site. In most hydrogeological studies, this will involve examination of three different data sets, commonly on different scales. Fracture zones are identified as lineaments on air photos or enhanced satellite imagery. Lithologies are determined from published geological maps or, in more remote areas, by reconnaissance ground mapping. The topographical elevation is determined from a contour map. It is of great advantage to the hydrogeologist to have these three critical data sets integrated in a single visually interpretable product.

Fracture zones, whether resulting from faulting or shearing, or from solution enlargement of joints or contacts between differing lithologies, are often good sites for boreholes for shallow (less than one hundred metres) groundwater extraction. They can also form channels for escape of polluted water from waste disposal sites. Their identification is thus of great importance in hydrogeology. The precise location of fractures is usually achieved by geophysical means, using geoelectrical, electromagnetic or seismic techniques, but their initial identification has for many years been carried out primarily by remote sensing .

As shown diagrammatically in Figure 8.1, fracture zones in sub-outcropping rocks can result in topographical irregularities and vegetation changes which permit their detection by remote sensing. Linear features detected as a result of vegetation, soil or topographical contrast, and extending over distances ranging from a few hundred metres to tens of kilometres, are often visible in air photos and remotely sensed imagery, and are known as lineaments. Not all lineaments are the result of fracturing. Some result from geological features of no interest to the hydrogeologist, others from relatively superficial processes such as glaciation, and many from human activities. The discrimination of hydrogeologically significant lineaments from the more numerous spurious lineaments depends to a large extent on the local knowledge and experience of the interpreter.

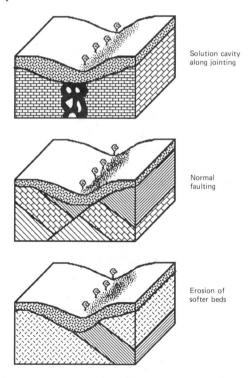

Figure 8.1. The relationship between geological structure and surface lineaments, as expressed by vegetation.

Air photos have been in regular use in hydrogeology for at least the past thirty years, and acceptance of satellite remote sensing by hydrogeologists, particularly those working in tropical and semi-arid environments, followed rapidly on the launch of the first Landsat earth observation satellite. The Landsat MSS, with its 80 metre spatial resolution, only permitted lineament interpretation at scales of 1:100,000 or smaller, and was therefore used mainly in the preliminary stages of a groundwater exploration programme, followed by detailed lineament interpretation on air photos. The organisation of a typical exploration programme in the mid-1970's is illustrated in Figure 8.2. MSS false-colour composites at a scale of 1:250,000 were used to define major lineaments, which were then examined in detail on 1:20,000 monochrome air photos prior to ground geophysical investigation and test drilling. The advent of finer spatial resolution scanners such as the Landsat Thematic Mapper (TM), (30 metres) and SPOT (10 metres in Panchromatic mode) permits lineament interpretation at larger scales of between 1:50,000 and 1:20,000, and could eliminate the need for intermediate air-photo interpretation before ground geophysical studies.

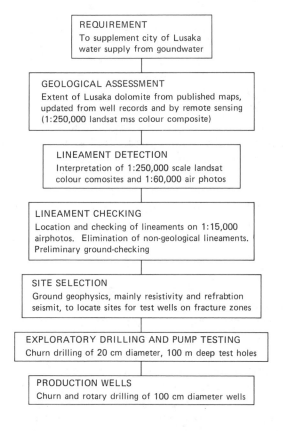

Figure 8.2. The organisation of a typical groundwater exploration campaign in the mid-1970's, showing the place of remote sensing.

Lineaments can be identified visually by a human interpreter, or automatically using suitably designed algorithms which take full advantage of the digital nature of the imagery. Lineament detection software is still in an experimental stage in that it can certainly detect all lineaments with an objectivity impossible in a human observer, but cannot emulate the experienced interpreter who eliminates many obviously spurious lineaments at an early stage of interpretation. The number of lineaments resulting from most automatic line and edge detectors is normally very large unless thresholds are set high; this introduces the risk of suppressing significant lineaments. Work is in progress to improve automated techniques, and their use will ultimately reduce the cost of lineament detection, as well as minimising subjective bias. For the present, however, visual interpretation is still the most reliable technique for lineament identification and assessment.

Satellite imagery can be enhanced in various ways to make visual interpretation of lineaments simpler and more reliable. In some cases lineaments are more clearly visible on colour composite images than in monochrome single-band products. Edge-enhancement algorithms can be used to accentuate edges and other linear features.

Different wavelengths and sensor systems do not always detect the same lineaments. Lineaments resulting from vegetation or soil contrast are usually most clearly seen in the reflected infrared wavebands such as TM bands 4 or 5. SPOT Panchromatic imagery, despite its excellent spatial resolution, does not discriminate well between different vegetation types, and is thus not ideally suited for lineament analysis. Imagery acquired by active microwave systems highlights variations in surface roughness, and in surface aspect relative to the sensor. The relatively low angle of most SAR systems results in an effect similar to low solar illumination angles in winter, highlighting minor topographical irregularities and thus emphasising lineaments with topographic expression. The problem with a radar image acquired from a fixed direction is that all lineaments parallel or sub-parallel to the satellite orbit will produce strong signatures, while those at right angles will be suppressed. The same directional bias is apparent in optical imagery acquired under low sun angles in winter. Where data sources and cost permit, it is desirable to employ a range of imagery from different sensor systems, in particular microwave and optical, to ensure detection of all significant lineaments.

Many hydrogeologically significant fracture zones allow groundwater leakage into overlying soils. The presence of unusually high concentrations of soil moisture can result in lowered soil temperatures, either because the groundwater is significantly cooler than ambient surface temperatures, or through evaporation in the capillary zone. It should theoretically be possible to detect such zones of anomalously lower temperatures using spaceborne thermal sensors. In practice, variations in thermal emission from natural land surfaces are controlled as much by their aspect in relation to incoming solar radiation, and their thermal emissivity, as by the heat flow from underlying soils. A thermal image acquired at a single time during the day, as is the case with Landsat TM, is of little use in locating zones of high soil moisture content. Night-time thermal imagery, and day - night thermal image pairs, are only routinely acquired with the Advanced Very High Resolution Radiometer (AVHRR) sensor on the NOAA TIROS-N satellites. This has a nadir resolution of 1.1 km, too coarse for most hydrogeological applications. Airborne thermal scanners can be used to aquire pre-dawn imagery for hydrogeological studies.

8.1.1. Data Integration for Fracture Analysis
A hydrogeologist will commonly make an interpretation of fractures from air photos or satellite imagery, and then evaluate the significance of these fractures by reference to a geological map. Intense fracturing in some rock units may be of much greater significance than in others. A fractured crystalline carbonate rock, for example, is likely to be a better aquifer than a fractured granite. In selection of actual sites for production boreholes, topography will also be important, since although the ground water table tends to follow surface topography, it does so in a smoothed fashion and is therefore further from the surface in areas of high elevation. Boreholes in topographical lows can usually be less deep than those drilled on hilltops, and will often provide higher yields because of lateral flow from higher areas. The combination of a satellite image highlighting fractures with a geological map and a digital elevation model should present the hydrogeologist with much of the information required for preliminary borehole site selection within a single map.

As an example of the possibilities of remote sensing and data integration in hydrogeology, Landsat TM band 4 imagery of the Isle of Wight, acquired in summer, was filtered using three mutually perpendicular 3*3 linear gradient filters, and the results of the three filters combined into a single image. The filtered image was then combined with a digitised geological map and with a digital elevation model, both in raster format. The combination was carried out using an intensity - hue - saturation (IHS) transform. Filtered satellite imagery was assigned to intensity, the digital geological map to hue, and the digital elevation model to saturation. A reverse colour transform to red - green - blue colour space was then carried out for the whole of the Isle of Wight and for a detailed extract centred on the town of Newport. The process is illustrated in Figure 2.10, and an example of the resulting composite image, combining filtered satellite imagery with geology and elevation, is shown in Plate 6 .

The cost of producing a product of this type for a large groundwater exploration programme, covered by a quarter TM scene (approximately 90*90 kilometres) is summarised in Table 8.1. This represents a total cost per square kilometre of only 0.63 UK pounds, which can be compared with the purchase price of air photos from archive of 1.60 per square kilometre, and an acquisition cost of new air photography in excess of five pounds per square kilometre. The additional convenience, and probably improved interpretability, of a colour composite product compared with use of separate air photos, geological and topographic maps is not easily quantifiable, but the cost-benefit of using remote sensing is apparent even in the raw data costs.

Table 8.1. Cost estimates for composite image of 90*90 kilometre area

Item	Cost in UK Pounds (1991)
Image acquisition	1100
Geometric correction	550
Purchase of DEM's	625
Purchase of geological maps	90
Digitisation of geological maps	1000
Image processing	300
Film writing	250
1:50,000 scale prints	500
SUB-TOTAL	**4415**
Administrative charges etc.(15%)	662
TOTAL	**5077**

8.2 DAM SITE AND PIPELINE STUDIES

The uses of remote sensing in evaluation of potential dam sites include many of the activities discussed in previous chapters, and in the first half of this chapter. The actual site of the dam should be in a stable area, free of major earthquakes, landslips and the likelihood of volcanic eruption, and should also not have a high density of fracturing. Certain lithological horizons might have a negative influence on the stability of the proposed dam, for example beds of montmorillonitic clays, graphite-rich schists, or even rocks unusually high in pyrite, and the mapping of·such units can in some cases be assisted by remote sensing.

Apart from the actual dam, consideration must be given to the area which will be inundated by the proposed reservoir. Existing land use in the area to be flooded can be mapped using remote sensing combined with ground observations. The amount of productive agricultural land which will be lost following dam construction can be measured, and estimates produced of the numbers of people likely to require re-settlement. Sites of special environmental significance - important wetlands, for example - can also be mapped using remote sensing. Historical sites may not be amenable to remote sensing study, but GIS techniques can make use of information from remote sensing and other sources, combined with a digital elevation model, to model the extent of flooding at different possible dam levels. A balance can then be sought between the maximum water level in the dam, and thus the amount of water stored, and the damage to sites of ecological and historical importance, and costs of re-settlement and loss of productive land.

In some special geological situations, the geology of the proposed lake bed can also be significant. The presence of major fracture zones in the area to be flooded can result in leakage once pressure is increased by filling the reservoir. A major hydro-electric storage dam in Zambia was sited on a deep-seated fracture zone, and an old hot-spring system just downstream from the dam was re-activated once the reservoir was flooded. Water percolated down into the fracture zone from the floor of the reservoir, and emerged below the dam heated and somewhat mineralised by its deep circulation. A series of expensive relief wells had to be drilled to reduce the water pressure in fracture zones beneath the dam, which might have threatened the stability of the large rock-fill structure. Fracture zones on this scale are very prominent features in remotely sensed imagery, and could have been located at the planning stage.

The siting of pipelines must combine geotechnical considerations with assessement of environmental impact. The route of the proposed pipeline should avoid areas of serious natural hazards such as active fault zones and land subject to landslips or volcanic eruptions, and should also minimise the aesthetic impact and the risk to the environment in the event of leakage, although except in delicate Arctic regions leakage of water is rarely dangerous to the environment. As discussed in the chapter on environmental impact assessment, remote sensing and GIS techniques have their part to play in choosing between different options with pipelines as with roads and railways, and allowing the planner to interactively compare a wide range of different possible options.

9
Applications of Remote Sensing During Production

Satellite remote sensing is normally thought of as a technique to be employed during geological mapping and mineral exploration, and possibly also at the mine-planning stage, but little thought has been given to applications during production. The relatively coarse resolution of most satellite imagery and the small size of some mine sites are certainly limitations on the use of remote sensing, but there are still areas in which the technology can be profitably employed.

9.1 UPDATING MINING DISTRICT LAND-COVER MAPS

Remote sensing does not yet provide sufficient spatial resolution to compete with conventional ground surveying or photogrammetry in the production of regularly updated maps of individual open-cast mines. What it can provide, however, is an overview of a large area, such as a whole mining district, in a digital and multispectral form. A single mining district may include the mine workings of a large number of separate, and often competing, mine operators. Individual mine maps are often produced at different scales, and using different classifications and keys. It is often necessary for regularly updated maps of the whole district, showing changes, new pits and dumps, restoration and plans for new developments, to be produced for local and national governments. Remote sensing can play an important role in this respect, since the resolution is appropriate to the production of final maps at scales of between 1:25,000 and 1:100,000, and the digital nature of the imagery permits objective detection of change. The china clay mining districts of the southwest of England are taken as an example of how remote sensing can be used in the production of regularly updated mining district maps, and in the detection of change.

China clay is Britain's most valuable mineral export after oil, the most extensive workings being in the St. Austell district of Cornwall. Hydrothermally altered granite is excavated by mainly hydraulic means, using powerful water jets or monitors, in large open pits. After some initial separation of large unweathered boulders in the pit, the resulting slurry of kaolin, quartz, muscovite and partially weathered feldspar is pumped out of the pits for further processing. A series of gravity separations results in production of a pure kaolinite slurry, which is dried in kilns. A residual slurry of fine-grained muscovite with minor kaolinite is disposed of in "mica dams", and a sand-sized waste product of quartz with minor feldspar, together with coarse rubble of partially decomposed granite, known in the mining areas as "stent", builds the large waste tips so characteristic of the china clay districts. Old spoil heaps were normally conical, resulting from transport of waste by small skips on fixed rails, but new dumps are much larger and usually flat-topped, since material is deposited mainly by dump trucks and conveyors. Old spoil heaps have become naturally vegetated with heather, grass and rhododendron thickets, while the sides of new flat-topped heaps are commonly grassed to prevent erosion and lessen their environmental impact.

The main objective of this study was to develop a technique for rapid low-cost production of thematic maps of the mineral district, on a regular basis, using available imagery irrespective of season. Detailed studies of the Lee Moor workings in Devon indicated that conventional classification techniques (box, maximum likelihood and unsupervised) did not produce reliable results, particularly at low solar elevations, with a large percentage of unclassified surfaces. Within the china clay workings, four spectrally distinct types of surface have been identified, as illustrated in Figure 9.1. Water in old and active pits is often extremely turbid, with high reflectance in the three visible TM bands but low reflectance in the infrared. "Mica" (mainly muscovite, with subordinate lepidolite and fine-grained kaolin) has a relatively high reflectance in band 4, with low reflectance in bands 5 and 7. "Clay" (mainly kaolinite, with subordinate feldspar and quartz) has roughly equal reflectance in bands 4 and 5, with a marked decrease in band 7. "Sand" (mainly quartz with subordinate feldspar) has roughly equal reflectance in bands 4 and 7, and markedly higher values in band 5. Studies of the flat and mineralogically simple surfaces of mica dams indicated that the use of band ratios largely eliminated the effects of solar elevation, remaining differences between different image dates due to atmospheric effects being minor in relation to the variation in response between different types of surface (see Figure 2.6 which illustrates the use of band ratios).

Digital maps from different dates can be used to measure change in areas of mineral workings. As an example, comparison of image maps for September 1985 and May 1988 indicates that new mineral working areas, mainly sand dumps, developed during that period occupied a total area of 223 hectares, while restoration of a total of 248 hectares of mining sites was achieved during the same period, indicating a net reduction of 25 hectares in the total area interpreted as mineral workings (Plate 11). Areas of rapid change, for example new sand dumps, are highlighted by comparison of image maps and can be studied in detail using co-registered image extracts, as shown in Figure 9.2.

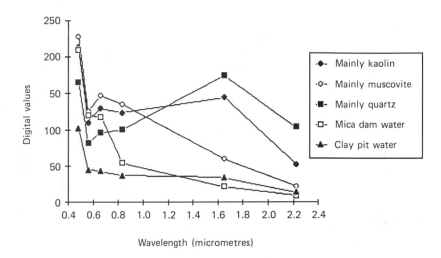

Figure 9.1. Digital values for mineralogically different cover types, Lee Moor china clay workings.

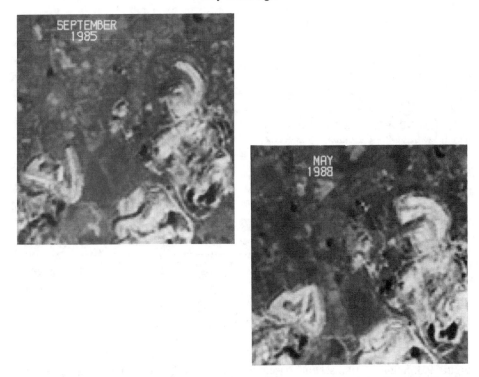

Figure 9.2. Co-registered extracts of Landsat TM imagery, part of the St. Austell china clay workings. The change in area of the dumps can be clearly seen.

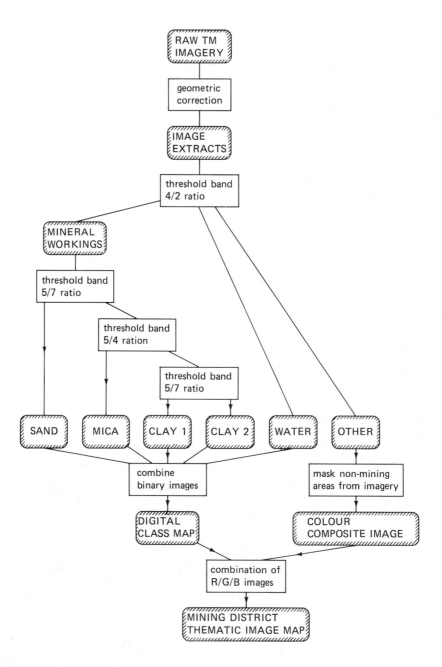

Figure 9.3. Segmentation scheme based on band ratios. Used for preparing land-cover maps of china clay mining districts.

Different mineralogical species exposed in the china clay workings are spectrally distinct, and they should be separable with TM imagery. If there is a requirement for regular mapping of this or similar areas using remote sensing, it is essential that the mapping technique should be as insensitive as possible to variations in solar elevation angle. The chance of obtaining imagery each year, or even each second year, at the same season and thus at similar solar illumination angles, is extremely remote. The success of band ratios in eliminating most of the seasonal illumination variation, at least on the horizontal surfaces of the mica dam, suggests that such ratios, carefully selected to be discriminatory of different surface types, could be used to segment and classify imagery.

The first requirement, if such a segmentation scheme is to succeed, is to separate mining areas from the surrounding land, whether it be moorland, woodland or arable. Experimentation has shown that the 4/2 ratio, shown to be extremely stable at different solar illumination angles, is consistently low for mining areas and high for all vegetated land. Very low values of this ratio, a kind of vegetation index, are seen over water. An initial segmentation is therefore made into water (lowest values of the ratio), mining areas (intermediate values) and other land (high), as shown in Figure 9.3. This is followed by a series of thresholding operations of different ratios to progressively segment the mining areas. A 4/7 ratio is first applied to separate sand-dominated areas. Absorption in band 7 is much less for quartz than for other mineral species, while the absorption in band 4 is greater (Figure 9.1). Low values of the 4/7 ratio for mining areas can thus be used to separate quartz-dominated surfaces. The 5/4 ratio is then used to separate mica from clay. Mica has the highest reflectance of the mineral species in band 4, and the lowest in band 5, so that low values of this ratio are diagnostic of mica-dominated surfaces. The 5/7 ratio is finally used to subdivide areas dominated by clay. Outcrops of partially kaolinised granite and dumps containing significant proportions of feldspar ("stent") are separated from purer kaolinite areas by means of this ratio. The actual threshold values for each ratio were established interactively for one date of imagery with reference to maps of surface cover in the mining areas, and were then applied to other dates of imagery. The definition of total mining areas was found to be very accurate (better than 90 %), as was the separation of water. The accuracy of the other categories has proved difficult to check, since the mine operators do not record the distribution of different mineral species on dumps and in pits, and a waste rock dump can be spectrally identical to a pit with outcrops of partially decomposed granite. The sequential thresholding process is, however, very rapid and can be carried out objectively with a minimum of operator intervention to produce an updated image map. The study demonstrated the feasibility of preparing, in a semi-automatic and rapid fashion, thematic maps subdividing china clay mining districts into six broad land-cover classes.

9.2 PRODUCTION AND EXPLORATION INTELLIGENCE

The need for accurate information about the activities of competitors is as vital in the minerals industry as in most other industries, and the methods of obtaining this information are diverse, extending from simple to very complex and from innocuous to extremely dubious in moral and legal terms. Not only mining companies, but also producer nations, require information about the performance and plans of other producers, particularly in the case of commodities having great strategic importance, or whose price is subject to extreme variations due to market forces or sentiment. In the gold industry, for example, a company planning a major new mine in South Africa, with the huge investment and long payback period associated with that investment which any large new deep mine implies, would need to be assured that no huge increase in production could be expected from elsewhere in the world. A sudden large expansion of Russian gold production, combined with increased foreign sales of Russian gold, could threaten the stability of world gold prices. Although most western gold producers, requiring investor support, announce development plans far in advance, this has not so far been the case with Russian producers, and industrial espionage was resorted to in order to provide this vital information. Another example could be an exploration boom situation, where companies compete with each other for ground, and are often extremely secretive about their activities at some stages of the exploration, while being in some cases less than honest about their discoveries at later stages. Industrial espionage may be used to try to ascertain the activities of the competition, and even to gauge their exploration success or lack of it.

Although present day commercial satellites do not have the spatial resolution to count the number of drill rods or core boxes next to a drill rig in order to assess the depth reached, or even to count the number of trucks at work in a large open-cast, they can provide valuable intelligence to the minerals industry, in a completely legal fashion. SPOT Panchromatic imagery, with its spatial resolution of ten metres, permits estimates of areas of open pits and dumps to within plus or minus about two hundred square metres, while stereographic SPOT PAN imagery, with its height accuracy of about fifteen metres, allows volumes of pits and dumps to be estimated to within three thousand cubic metres. Yearly observations of large mines, combined with other information about average grades, can give acceptable estimates of the annual production. The use of multispectral imagery, such as that from TM, in combination with SPOT PAN imagery, can give accurate information on surface disturbance associated with pre-mining stripping operations, engineering works associated with new processing plants, or construction of new transport routes. Remote sensing will rarely totally supplant other sources of information, but can provide both an early warning that changes are likely to occur and supporting visual evidence of changes indicated by other sources. In the exploration field, the activities of mineral exploration teams in remote areas are often only too visible from space. Cut lines for geophysical and geochemical surveys, and access roads to drilling sites, can often be observed on satellite images many years after the completion of exploration. Individual drill sites are easily recognisable, often by the unchecked pools of slurry, and the pattern of a newly discovered mineral deposit is often revealed by the distribution of drill sites. No examples are known of the recognition of rock and mineralisation types in drill slurry using remote sensing, but the many (possibly apocryphal) stories about the use of light aircraft to monitor the colour of drill slurry during the western Australian nickel boom suggests at least the possibility of remote sensing having a role here.

As an example of the way in which remote sensing can be used to provide information about changes in operating practice in a large mine, we can look at the Chuquicumata copper mine in Chile. Fine-resolution imagery from the SPOT satellite has only been available since 1986, and changes since that date are likely to have been relatively minor. Landsat MSS imagery has been available since the early 1970's, and despite its coarse spatial resolution, this can be compared with recent SPOT imagery to detect longer term changes. The mine as it was in 1987 is shown on the right of Plate 16, and an MSS image acquired in 1974 appears on the left of the same plate. The most obvious changes during the thirteen year period are the considerable expansion of waste dumps. For much of its life, this very large mine has operated at very low stripping ratios, since the shape of the porphyry copper orebody has been such that almost all the material mined from the open pit has been of ore grade. Deepening of the pit over the past twenty years has required the excavation of large volumes of waste rock from around the edges of the pit in order to ensure stability of the steep slopes, and this has resulted in a great increase in waste rock dumps. Stereo imagery would permit estimation of dump heights, and thus of their volume, but even the use of a single image can permit a crude estimate of heights of features steep enough to cast a shadow, since the solar elevation at the time of image acquisition is known. The principle of this method of height estimation is illustrated in Figure 9.4, a PXS (SPOT PAN and XS imagery acquired simultaneously and combined by SPOTIMAGE) image of central Johannesburg. The shadows of tall buildings can be clearly seen, and their length measured. Simple trigonometry permits estimation of the heights, since the solar elevation angle is known.

Using this technique, the volume increase in waste dumps at Chuquicumata between 1974 and 1987 could be estimated. Detailed knowledge of the mine from other sources would allow an investigator to estimate total production, or the change in depth of the open pit, from this information. Reference to Plate 16 also indicates that the volume of slag has increased significantly, as might be expected, and it appears that a series of tanks, possibly a new leach plant, have been installed next to the concentrator.

Figure 9.4. A PXS image of part of central Johannesburg showing the shadows of tall buildings which can be used to estimate their heights (Copyright CNES 1988).

10
Land Restoration

Once mining is complete, there is commonly a requirement to restore the land surface to its pre-mining use, which in populated areas usually means restoration to agricultural land. It is desirable to achieve similar agricultural productivity on restored land to land never mined, and there is an increasing requirement for the scenic quality of the land to be retained. There may even be opportunities for scenic improvement during restoration, with the addition, for example, of small hills in a previously featureless landscape, or the provision of ponds and small lakes. Restoration planning may involve conflict between farmers, who wish to maximise productivity by re-planning the field layout, replacing original small irregular fields by large rectangular fields, and conservationists, who would like to retain the network of hedgerows in the old field pattern.

10.1 PRE-PRODUCTION LAND-USE STUDIES

If restoration of land after mining is to return the land as closely as possible to its original state, it is important that as detailed a record as possible of the pre-mining state of the land is obtained before any disturbance occurs. Maps should be prepared of soil types and thicknesses, and of course of the topography. Remote sensing cannot usually help in acquisition of this type of data, but it can permit a fairly detailed assessment of the distribution of different types of land use and land cover before mining. Land-use maps, prepared as described in previous chapters, can be used as the baseline against which restoration can be planned and monitored. Apart from mapping the main land-use categories, remote sensing can provide generalised but quantitative information on vegetation vigour at different seasons by means of vegetation indices. Pre-mining vegetation indices, measured at critical times in the crop growing season, probably using airborne scanners since acquisition of imagery at specific dates is still difficult using satellites, can be used as the reference against which post-restoration land productivity can be measured.

Satellite imagery acquired before the start of mining can be combined with digital elevation models to generate perspective views of the site. These can then be used by the landcsape planners to model possible changes or improvements to be incorporated in the post-mining landscape. These perspective views can also be extremely valuable visual aids in public inquiries which may precede the granting of planning permission for mining.

10.2 MONITORING OF PRODUCTIVITY OF RESTORED LAND

Remote sensing can play an important and cost-effective role in assessing the quality of restored land, as well as in comparing the size and shapes of fields in restored and unmined areas. A remote sensing study of an area of open-cast coal mining in Northumberland in the north-east of England demonstrates the use of vegetation indices to compare productivity of restored and unmined land.

The coastal area of southern Northumberland has a long history of coal mining. Most production in the past has been from underground or "deep" mines, but only a few of these still operate. Open-cast production started in the early 1940's, but it was not until 1956 that large-scale open cast mining commenced. A total of thirteen open-casts have operated in the area since then, five being in production in the late 1980's (Figure 10.1). All mined areas are restored to agricultural or recreational use after mining. The main agricultural activity in the area is sheep grazing although minor acreages of grain crops, potatoes and oil-seed rape are grown. Most of the restored land is under pasture, with strips of woodland to act as windbreaks and to improve the scenic quality. Restoration is usually a continuous process, concurrent with mining. Topsoil, subsoil and waste rock from the initial cut of an open pit are stored next to the site. As mining proceeds laterally from this initial cut, waste rock is moved, usually by dragline, from above the coal seams and dumped directly in the worked-out part of the pit. Topsoil and subsoil are removed from ahead of the advancing face of the pit, and placed directly on rock-filled mined-out areas. Once mining is complete, the final pit is backfilled with stored material from the initial cut. The process of restoration results in an homogenisation of soils and, where the ratio of coal to waste is small, may leave the restored land at a slightly higher mean final elevation than the original surface owing to imperfect compaction of waste rock. In cases where the ratio of coal to waste is high, the final cut has been inadequately filled with material at the end of mining, resulting in a lower than average land surface and even some open water-filled voids. Success of restoration is reported to have increased as the mine operators and British Coal gained experience. More consideration has been given to aesthetics in recent restoration, with trees being planted in less geometric patterns, and excess backfill material being used to produce undulating surfaces. Drainage can be a problem, particularly in low-lying coastal sites. Excessive compaction of the sub-soil may occur due to heavy machinery used for transporting and dumping the soil, reducing soil porosity and sometimes causing waterlogging of topsoil. Most restored areas are planted with grass, and the farms are ultimately either handed back to their original owners, or if purchased prior to mining by British Coal, sold to new farmers.

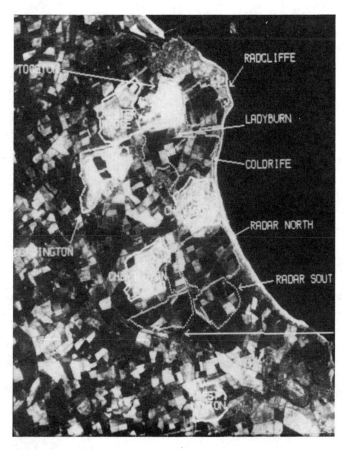

Figure 10.1. Open-cast mining sites in the Druridge Bay area, Northumberland, United Kingdom.

Landsat MSS imagery from 1973, 1976, 1977, 1979 and 1981 was co-registered with TM imagery for 1984, 1985 and 1986, and images of the near infrared to red ratio (MSS bands 7/5, TM bands 4/3) were prepared for each date. Values of this ratio were extracted for three coal mine sites, restored to agriculture in the mid 1970's, for all dates of imagery, and were compared with ratios from three control areas of similar agriculture on unmined land. Mined and unmined sites were all in excess of one thousand pixels in area, and a crude atmospheric correction was applied to the near infrared / red ratios based on the assumption that, for each date, the value of this ratio over open sea should be zero. No corrections were applied for solar elevation since the use of ratios is assumed to largely eliminate illumination differences. The mean ratios for mined and unmined land were then plotted on a graph of time against the near infrared / red ratio (Figure 10.2), where time is the Julian day of image acquisition, irrespective of year.

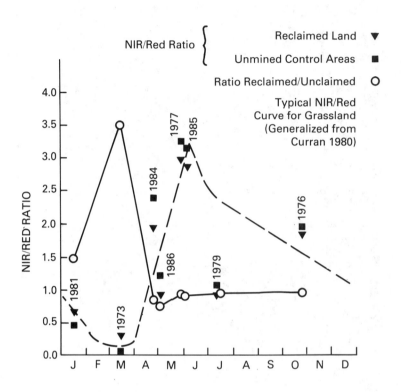

Figure 10.2. Comparison between vegetation indices for restored and unmined land, Druridge Bay, Northumberland.

With the exception of the July (1979) scene, the ratios lie very close to a generalised curve of grassland vegetation index. During spring, summer and autumn the ratios for restored land are slightly lower than for unmined land, the difference being greatest in the spring. In winter, on the other hand, restored land ratios are higher than those for unmined land. This study showed that, in general, pasture on restored land has almost the same vigour as that on unmined land; the larger differences in spring suggest that grass on restored land starts its most active period of spring growth after unmined land. This could be due to lower porosities of restored land, resulting from compaction by heavy machinery during restoration. Quantitative information of this type is not obtainable by air photography, and would require expensive field work at many different dates to collect on the ground.

10.3 POST-PRODUCTION LANDSCAPE QUALITY ASSESSMENT

Landscape quality is an extremely subjective concept, and may mean totally different things to different observers. A person who has devoted his life to building up a massive industrial enterprise may regard a landscape of factories and smoking chimneys as having great beauty, while the same landscape may be totally abhorrent to a farmer. Any landscape contains aesthetic and economic elements, and the weight given to each depends on the inclinations, background and objectives of the observer.

In the specific case of land restoration after mining in predominantly agricultural areas, there are certain criteria by which the succes of restoration is usually judged. Most observers would agree that the land should look at least similar to surrounding areas of unmined land, and that the agricultural productivity of the land should not be impaired by mining. Economic considerations may, however, result in landscape changes after mining. Most areas of the world that have been settled and farmed for generations have extremely complex patterns of field boundaries, resulting from sub-division of land within and between families, and determined to some extent by inheritance practices of the country. These irregular fields, often with extensive hedgerows, may result in a very varied habitat for plants, birds and animals, but they are rarely the most profitable way of farming. Mining, and the land restoration which follows it, offers the possibility of re-modelling the landscape in a more practical form. Original small irregular fields can be replaced by larger more uniform fields which are cheaper to plant and harvest. The farmers benefit, as does the mining organisation, since this type of restoration is significantly cheaper than replacing the complex pattern of hedgerows. An improvement in landscape quality, from the point of view of the farmer, but a reduction in landscape quality for the diverse species inhabiting the hedgerows. The other negative result of this type of restoration is also that the restored mine sites remain clearly visible as patterns of large regular fields amongst a maze of old irregular fields.

The impressions created by such a landscape are largely subjective, but it is possible to quantify such aspects of landscape pattern as field size and shape. Image processing techniques can be used to extract fields of different crops, and then to analyse their size distribution and boundary lengths. An objective comparison can then be made between areas of traditional farming land and the restored areas with more rational field boundaries. Statistics of this type can obviously be used in many ways, and be subject to widely differing interpretations, but it is surely a better basis for discussion to be able to measure something quantitatively than to describe it in purely subjective terms.

The spatial association of different types of land cover is another important aspect of landscape quality. A landscape combining groves of trees, fields of wheat, maize or rice, grassland or pasture, and streams and lakes, would usually be considered more satisfying than one which was either all forest or all crops. Diversity is thus important, and can be measured using GIS techniques. The ecological quality of a landscape may depend on the range of habitats which are available, and thus also on diversity, but the suitability of a landscape for a given species may also depend on the spatial association of, for example, woodland and open water. The length of woodland edges against farmland is an important environmental indicator for some species assemblages, a highly irregular pattern of woodland being preferable to a large single rectangular forest. These parameters can again be quantified using remote sensing and GIS's. The whole field of pattern and shape analysis is one which is still in its infancy in remote sensing, but is likely to grow and mature over the coming decade.

Bibliography

GENERAL REMOTE SENSING TEXTS
American Society of Photogrammetry, **Manual of Photographic Interpretation**; Falls Church, Virginia, 1960.
American Society of Photogrammetry, **Manual of Remote Sensing**, 2nd ed.; Falls Church, Virginia, USA, 1983.
Lillesand T. M. and Kiefer R. W., **Remote Sensing and Image Interpretation**; 2nd ed. John Wiley, New York, 1987.
Ray, R. G., **Aerial Photographs in Geologic Interpretation and Mapping**; USGS Professional Paper 373, U. S. Govt. Printing Office, 1960.
Sabins, F. F. jr., **Remote Sensing: Principles and Interpretation**, 2nd ed.; Freeman, New York, 1986.

SENSORS AND SATELLITES
Allan, T. D. (editor), **Satellite Microwave Remote Sensing;** Ellis Horwood, Chichester, 1983.
Evans, D. L., Farr, T. G., Ford, J. P., Thompson, T. W. and Werner, C. L., **Multipolarisation Radar Images for Geologic Mapping and Vegetation Discrimination**; IEEE Transactions on Geoscience and Remote Sensing, vol. GE-24, no. 2, March 1986, pp. 246-257.
Kidwell, K. B., **NOAA Polar Orbiter data users guide**; NOAA Satellite Data Services Division, World Weather Building, Washington, DC. 1983
Legg, C. A., **Digital Satellite Imagery from the Soviet Union; An Important New Tool for Environmental Monitoring;** Procds. Remote Sensing Society Conference, Swansea, 1990, pp. 65-74.
Legg, C. A., **A Review of Landsat MSS Image Acquisition over the United Kingdom, 1976-1988, and the Implications for Operational Remote Sensing**; International Journal of Remote Sensing, vol. 12, no.1, 1991, pp. 93-106.
Palluconi, F. D. and Meeks, G. R., **Themal Infrared Multispectral Scanner (TIMS): An Investigators Guide to TIMS Data**; NASA, JPL Publication 85-32, 1985.
Pease, C. B., **Satellite Imaging Instruments;** Ellis Horwood, Chichester, 1991.
Price, J. C., **Satellite Orbital Dynamics and Observation Strategies in Support of Agricultural Applications**; Photogrammetric Engineering and Remote Sensing, vol.48, no. 10, October 1982, pp. 1603-1611.
Ulaby, F. T., Moore, R. K., and Fung, A. K., **Microwave Remote Sensing, Active and Passive**; (3 volumes). Artec House, Dedham, Mass. 1981-1986.
USGS and NOAA. **Landsat 4 Data Users Handbook**; USGS, Sioux Falls, SD, 1984.

Vane, G. and Goetz, A. F. H., **Terrestrial Imaging Spectrometry**; Remote Sensing of Environment, vol. 24, 1988, pp. 1-29.

Wooding, M. G., **Imaging Radar Applications in Europe**; ESA/JRC. Esa Publication TM-01, October 1988.

ARCHIVING AND DISTRIBUTION

Slogget, D. R., **Satellite Data: Processing, Archiving and Dissemination;** Ellis Horwood, Chichester, 1989.

IMAGE PROCESSING

Fabbri, A. G., **Image Processing of Geological Data**; Nostrand Reinhold, Victoria, 1984.

Gillespie, A. R., Kahle, A. B. and Walker, R. E., **Color Enhancement of Highly Correlated Images**; Remote Sensing of Environment, vols 20 and 22, 1986 and 1987, pp. 209-235 and 343-365.

Jensen, J. R., **Introductory Digital Image Processing: A Remote Sensing Perspective**; Prentice-Hall, NJ. 1986.

Schowengerdt, R. A., **Techniques for Image Processing and Classification in Remote Sensing**; Academic Press, New York, 1983.

GEOLOGICAL MAPPING AND MINERAL EXPLORATION

Drury, S. A., **Image Interpretation in Geology**; Allen and Unwin, London, 1987.

Goetz, A. H. et al, **Imaging Spectrometry for Earth Remote Sensing**; Science, vol. 228, no. 4704, June 1985, pp. 1147-1153.

Harding, A. E. and Forrest, M., **Analysis of Multiple Datasets: Operational Advantages of an Automated Approach**; Remote Sensing Society Conference Proceedings, Bristol, 1989.

Institution of Mining and Metallurgy, **Remote Sensing: an Operational Technology for the Mining and Petroleum Industries**; Conference proceedings, London, UK, 1990.

Kahle, A. B. and Goetz, A. H., **Mineralogic Information from a New Airborne Themal Infrared Multispectral Scanner**; Science, vol. 222, No.4619, October 1983, pp. 24-27.

Loughlin, W. P., **The use of Principal Components Transforms in Mapping Hydrothermal Alteration Zones**; Proceedings ERIM Conference, Denver, 1991.

Siegal, B. S. and Gillespie, A. R., **Remote Sensing in Geology**; Wiley, New York, 1980.

MINING

Irons, J. r. and Kennard, R. L., **The utility of Thematic Mapper Sensor Characteristics for Surface Mine Monitoring**; Photogrammetric Engineering and Remote Sensing, vol. 48, no. 4, March 1986, pp. 389-396.

Legg, C. A., **Updating Thematic Maps of Mining Districts;** Procs Remote Sensing Society Conference, Bristol, 1989, pp. 243-248.

Appendix

Common Acronyms in

Remote Sensing

AIS	Airborne Imaging Spectrometer
AMI	Active Microwave Instrument (on ERS-1)
APT	Automatic Picture Transmission (for AVHRR, METEOSAT, etc.)
ATM	Airborne Thematic Mapper (Daedelus scanner)
ATSR	Along-Track Scanning Radiometer (ERS-1)
AVHRR	Advanced Very High Resolution Radiometer
AVIRIS	Airborne Visible and Infrared Imaging Spectrometer
BER	Bit Error Rate
BGS	British Geological Survey
CAD	Computer Aided Design
CCD	Charge-Coupled Device
CCT	Computer Compatible Tape
DEM	Digital Elevation Model
DTM	Digital Terrain Model (synonymous with DEM)
ERS	European Remote sensing Satellite
ESA	European Space Agency
EUMETSAT	EUropean METeorological SATellite organisation
GIS	Geographic Information System
HDDT	High Density Data Tape
HRV	High Resolution Visible (for SPOT sensors)
IFOV	Instantaneous Field of View
IHS	Intensity - Hue - Saturation (colour space or colour transforms)
IRS	Indian Remote sensing Satellite

JERS   Japanese Earth Resources Satellite
LISS   Linear Imaging Self-scanning Sensor (on IRS-1)
LUT   Look-Up Table
MOS   Marine Observation Satellite
MSS   MultiSpectral Scanner
NASA   National Aeronautical and Space Administration (USA)
NASDA   National Aeronautical and Space Development Agency (Japan)
NDVI   Normalised Difference Vegetation Index
NOAA   National Oceanographic and Atmospheric Administration
PDUS   Primary Data User Station (Meteosat data format)
PXS   Panchromatic and XS SPOT imagery, merged by SPOTIMAGE
RGB   Red - Green - Blue (colour space or display)
RIM   Regression Intersection Method (of atmospheric correction)
SAR   Synthetic Aperture Radar
SLAR   Side Looking Airborne Radar
TIMS   Thermal Infrared Multispectral Scanner
SIR-A   Shuttle Imaging Radar - A
SIR-B   Shuttle Imaging Radar - B
SNR   Signal to Noise Ratio
SPOT   Satellite Pour Observation de la Terre
TM   Thematic Mapper

Index